Physical Design of CMOS Integrated Circuits Using L-EDIT™

John P. Uyemura
Georgia Institute of Technology

PWS Publishing Company

I︎(T)P   An International Thomson Publishing Company

Boston • Albany • Bonn • Cincinnati • Detroit • London • Madrid • Melbourne • Mexico City
New York • Pacific Grove • Paris • San Francisco • Singapore • Tokyo • Toronto • Washington

PWS Publishing Company
20 Park Plaza, Boston, Massachusetts 02116-4324

I(T)P

International Thomson Publishing
The trademark ITP is used under license

For more information, contact:

PWS Publishing Co.
20 Park Plaza
Boston, MA 02116

International Thomson Publishing Europe
Berkshire House 168-173
High Holborn
London WCIV 7AA
England

Thomas Nelson Australia
102 Dodds Street
South Melbourne, 3205
Victoria, Australia

Nelson Canada
1120 Birchmount Road
Scarborough, Ontario
Canada MIK 5G4

International Thomson Editores
Campos Eliseos 385, Piso 7
Col. Polanco
11560 Mexico D.F., Mexico

International Thomson Publishing GmbH
Konigswinterer Strasse 418
53227 Bonn, Germany

International Thomson Publishing Asia
221 Henderson Road
#05-10 Henderson Building
Singapore 0315

International Thomson Publishing Japan
Hirakawacho Kyowa Building, 31
2-2-1 Hirakawacho
Chiyoda-ku, Tokyo 102
Japan

Library of Congress Cataloging-in-Publication Data

Uyemura, John P. (John Paul)
    Physical design of CMOS integrated circuits using L-edit / John P. Uyemura.
        p.   cm.
    Includes bibliographical references and index.
    ISBN 0-534-94326-8
    1. Metal oxide semiconductors, Complementary -- Design and
    I. Title.
    TK7871.99.M44U93   1995
    621.3815 -- dc20   94-28799
                                 CIP

Editorial Assistant: Cynthia Harris
Assistant Editor: Ken Morton
Acquisitions Editor: Tom Robbins
Production Editor: Helen Walden
Manufacturing Coordinator: Ellen Glisker
Marketing Manager: Nathan Wilbur

U.S. ISBN 0-534-94326-8
International ISBN 0-534-94327-1

Printed in the United States of America
95 96 97 98 -- 10 9 8 7 6 5 4 3 2

Dedication

This book is dedicated to my mother,
Ruby Shizue Uyemura.

植　村

Preface

CAD tools are critical for educating engineering students in the field of VLSI design. There are three main classes of tools that are required for designing integrated circuits: electronic simulators, layout programs, and logic simulators. At the electronics level, circuit simulators such as SPICE allow one to analyze the interaction of device parameters with the overall performance. Since SPICE variants are widely available on a variety of platforms, everyone can have access to this capability. The same is true for logic simulators, with many powerful programs available to the student. However, layout tools have not migrated into the realm of inexpensive software, and most existing programs were designed for workstations running a Unix environment. Historically, this was due to the fact that desktop PCs simply were not powerful enough to run the code. This, however, is no longer the case; state-of-the-art desktop computers have no problem handling many of the tasks that used to be the exclusive domain of workstations.

An important characteristic of an integrated circuit is that the electrical performance of the circuit is a sensitive function of the physical structure and layout. Chip design intrinsically links the electrical properties of a network with the physical characteristics of the patterned layers created during the fabrication process. The layout of integrated circuits thus becomes the central problem in high-density chip design.

This book is about using the L-Edit program, a product of Tanner Research, for designing integrated digital CMOS circuits. Included with the book is a special Student Version of the L-Edit program that runs on MS-DOS computers with only minimal hardware requirements. Although L-Edit can be used as a general drawing tool, the database files provided on the disk make it a powerful full-featured layout editor for CMOS integrated circuits. The procedures of design rule checking (DRC) and circuit extraction (which creates a SPICE-compatible netlist file from the layout) are available from within the program, establishing the critical link between physical design and circuit performance. Group operations, such as cell definition and instancing, can be used to create libraries for use in designing ASIC and other cell-based design techniques. And, the L-Edit program is very easy to learn and use.

Philosophy

In my opinion, it is not possible to understand the intricacies of integrated circuit design without actually creating the layouts, extracting the circuits, performing simulations, and then examining the interplay between the physical design and the electrical characteristics. This applies to both digital and analog circuits, from the transistor level up through the system level. Mastery of the subject requires that this process be repeated many times.

I have long advocated the use of CAD tools in the classroom. In 1988, I was invited to give a talk at the NSF-sponsored VLSI Educator's Symposium about the evolution of the microelectronics program at Georgia Tech. During the conference, it became obvious that many schools wanted to build a curriculum in VLSI, but found it a difficult task for one reason or another. A universal problem in academics is finding the means to acquire equipment. In the case of computer hardware, this directly affects what can be taught in the classroom. Even if workstations can be made available to the student, practical constraints often restrict the access to the machines, or the amount of time that each student can work. The latter comment can be particularly important when factoring in the learning time needed to become proficient with some of the more complex tools.

It was at this conference that I met John Tanner, and was introduced to an early version of the L-Edit program. During lunch he explained his view that the computing power of desktop PCs would quickly evolve to the point where professional level CAD tools could be supported. Over the years I have watched Tanner Research grow in both size and scope. When I contacted John in 1993 and proposed the release of a Student Version of L-Edit, he was quite enthusiastic about the project, and gave it his full support. This project is the result of my sharing John's view that the desktop PC has developed to the point where it can be used for real-world engineering.

Use of the Software

In this software, we offer a vehicle for learning chip design on a personal computer. When coupled with a SPICE simulator, most circuits encountered in an undergraduate or first-year graduate VLSI design course can be designed and simulated. It only takes a few minutes to learn the basics of using L-Edit. And, the advanced features such as the cross-sectional viewing program, the design-rule checker, and the circuit extraction algorithm provide the ability to perform complex designs.

I have discovered that L-Edit can serve as an excellent pedagogical tool if students are encouraged to try out their own designs. By performing the layout, studying various cross-sectional views, and simulating the circuits, important ideas are reinforced in real-time. In the classroom, I always stress the need for a critical analysis and re-design of a circuit. Since introducing the use of individual copies of L-Edit, I have noticed that the students tend to spend more time thinking about the critical parameters and modifying their circuits accordingly.

What about existing layout editors, such as Magic and commercial tools? Good question. At Georgia Tech, we are fortunate to have several workstation clusters that are available for student use. When I compared L-Edit on a PC with Unix-based layout editors, two advantages became clear. First, L-Edit is simple to learn and use. In teaching the program to my students, it seemed to take only 10-15 minutes until they were comfortable running the program and creating MOSFET layouts. The second advantage is that of portability. Many students own desktop computers, but only a select few can afford to have a workstation. Using L-Edit makes it viable to assign homework problems that are both complex and realistic, and not have worry about the students having access to a workstation. And, as you can verify, the performance of many L-Edit operations on a PC are often faster than equivalent commands in Magic on a Unix workstation. Moreover, since the full professional version of L-Edit is available on Unix platforms, one can transfer files from the Student Version to and from workstation clusters as needed. It is also worthwhile to mention that, in addition to L-Edit, Tanner Research offers a complete range of CAD tools in an integrated environment.

Use of the Book

When this project was first conceived, the main objective was to make the software available with a basic user's guide. While this was fine for those experienced in chip design, it seemed that a short tutorial on applying the program to CMOS design might be a worthwhile addition. In its final form, the book can be used either by itself, or in conjunction with a course in CMOS circuits and systems. Chapter 1 is an introduction to the installation and use of the L-Edit program. It illustrates many of the basic commands and operations needed to create and edit geometrical objects. It is recommended reading for all, regardless of background or experience.

The remainder of the book consists of two main parts. The first part, made up of Chapters 2 through 6, is an introduction to the design, layout, and analysis of digital CMOS integrated circuits. The treatment is quite general, but includes many direct references to the L-Edit program. For example, Chapter 4 is concerned with the theory and modelling of MOSFETs, but also shows in a step-by-step manner how MOSFETs are drawn in L-Edit with MOSIS. The writing is purposely tutorial in nature, and is intended to provide only a basic background in CMOS. Detailed calculations, special circuits, and advanced techniques were purposely omitted.

A few exercises have been included in the tutorial. For the most part, they are general design-type problems without any unique solution. The exercises were written to provide guidance for those pursuing individualized study, and to clarify some general concepts on using L-Edit in CMOS design. Since each school has a different set of constraints, the building of a local cell library seems to be the best approach to finding more relevant problems. Owing to this, no solution manual is planned.

The second portion of the book is a general reference on the L-Edit program, and is found in Chapters 7 through 11. Chapter 7 is a list of L-Edit commands, while the remaining chapters discuss the intricacies of advanced features. When combined with the basic information in Chapter 1, this material should provide sufficient information to use the program effectively and customize it as needed.

Acknowledgments

The staff of Tanner Research has been exceptionally generous with both time and energy, and were always willing to answer a question or look at a problem. Special thanks are due to John Tanner, Jim Lindauer, Anant Adke, Munir Bhatti, Bushan Mudbhary, Chris Yiu, and Mike Pottenger for their support and help in getting this project into final form.

Tom Robbins, my editor at PWS Publishing, has listened to me rant and rave about new book projects for the past 15 years. I am lucky that he continues to support these ideas without questioning my sanity (much). Helen Walden has been very helpful in getting this book into production, and Cynthia Harris never misses a detail, no matter how small. I am also grateful to Edward F. Murphy and John-Paul Lenney for their support in this project.

We were fortunate to have outstanding reviewers on this project. Professor Madeline Y. Andrawis, South Dakota State University, painstakingly went through the manuscript on a word-by-word basis (gritting her teeth at times, I'm sure). She spotted countless errors and made many valuable suggestions that helped to make the book more readable. Professor Donald W. Bouldin (University of Tennessee, Knoxville) and Professor Edgar Sanchez-Sinencio (Texas A&M University) also provided useful comments and insight into the project.

Thanks are due to the students in my 1994 courses, EE 6369 and EE 6370, for testing the software in their homeworks and design projects. They provided an endless source of energy by diligently working through complex layouts and providing many comments on the use of the program.

My wife, Melba, and my daughters, Valerie and Christine, provide the strength and motivation for taking on projects like this. Although I'll never be a millionaire, they always make me feel like one with their love and encouragement. I would also like to thank my parents, Reverend George and Ruby Uyemura, for their endless support and interest in my career.

Atlanta, Georgia
June, 1994

Contents

Chapter 1   Introduction to L-Edit

Chapter 6   Concepts for VLSI Chip Design

Chapter 10 Design Rule Checker

Chapter 11 Circuit Extraction Using L-Edit

PROPERTY OWNER NOTIFICATION QUESTIONNAIRE

1. Were you aware that a document affecting your property was recorded before you were notified by the Department of Registrar-Recorder / County Clerk?

   ☐ Yes   ☐ No

2. Do you believe it is a good procedure for the Department of Registrar-Recorder / County Clerk to notify parties when deeds and/or loans are recorded?

   ☐ Yes   ☐ No

3. Other Comments: _____

   _____

   _____

Optional
Name _____

Address _____

Phone Number (____) _____

Index

Chapter 1

Introduction to L-Edit

Welcome to the Student Version of L-Edit^{TM}! In this chapter, we will introduce you to the L-Edit program: how to install it onto your computer, what it can do, and how to use it.

1.1 Introduction

L-Edit is a product of Tanner Research, Inc.

L-Edit is an integrated circuit layout editor that was originally designed for desktop personal computers. It provides all of the features that are needed to design integrated circuits of arbitrary complexity using a generic MS-DOS PC with a minimal amount of hardware.

The Student Version of L-Edit contains all of the major tools provided with the full version, including circuit extraction, a design rule checker (DRC), and a cross-sectional viewer that allows you to see the layers in the finished chip. Every effort was made to ensure that this version of L-Edit would be useful in both classroom and research applications.

1.1.1 System Requirements

To install and run the Student Version of L-Edit, you will need a system with the following minimum specifications:

- An MS-DOS computer with at least 640K of system RAM and a high-density 3.5" floppy disk drive;
- A color monitor with EGA (or better) resolution and a compatible video card with a minimum of 256K RAM;
- A mouse or other pointing device, such as a trackball or a tablet.

Although L-Edit will run with the minimum system configuration listed above, the following system enhancements improve the performance:

- **A hard drive.** This will allow the program to run much faster.

- **A 3-button mouse.** L-Edit works the best with a 3-button mouse, such as those manufactured by Logitech.

- **Extended or expanded memory configured as a RAM disk.** Installing the L-Edit program onto a RAM disk is the easiest way to enhance the speed of L-Edit operations in this version.

Any or all of these system additions will improve the user interface with L-Edit. The program performs all calculations using integer operations, so it does not require (or use) a coprocessor. In addition, the Student Version will run under Windows or OS/2 as a DOS application[1].

L-Edit can provide hard-copy output using an Epson-compatible printer or an HPGL-compatible plotter. Postscript printers are also supported.

1.1.2 Limitations of the Student Version

Although the Student Version of L-Edit contains all of the drawing features of the full professional version, the following limits have been imposed:

- Memory usage is limited to the 640K system RAM, whereas the full professional version can access all of the available memory. This limits the size of the design files to about a maximum of around 50-60K.

- Data files can only be saved in .tdb form, which is a proprietary format of Tanner Research Inc.; "tdb" stands for Tanner data base. This does not affect printing, plotting, reading, or editing of the file.

- This version will not read, nor will it create, CIF or GDS II-format files. In particular, designs that are created with the Student Version cannot be directly submitted to an IC foundry for fabrication. A copy of the professional version is required to translate a .tdb file into these formats. The files produced with this software are fully compatible with L-Edit professional version 5.X.

- The Student Version is designed to produce EGA monitor resolution, while the full professional version also supports VGA and SVGA modes.

These limitations do not affect the operation of the software or the information contained in the data files.

Since this version works under the 640K DOS limit, it is very important that you do not use up system RAM with system utilities or other programs. Optimum performance is achieved when there are no resident programs loaded into your system. Even though the RAM access is limited, there is sufficient memory available for a typical MOSIS (MOS Implementation Service) Tiny Chip design.

The constraints on the Student Version were chosen in a manner that allows you full access to most of the features contained with the professional version of L-Edit without the cost. Tanner Research offers substantial academic discounts on the full professional version for schools that wish to translate designs generated by the Student Version software for fabrication. If you have very large chips, then the full version may be needed to assemble the cells from the limited-RAM version. (L-Edit

---

[1] Use the /w option on the L-Edit command line.

has been used to design chips with more than 500,000 transistors.)

Although the Student Version of L-Edit is only offered in an MS-DOS format, the full professional version is also available for Windows 3.1, and on Macintosh and UNIX platforms. Data files may be traded across different platforms.

1.2 Installing L-Edit

L-Edit may be run directly from a floppy disk or installed on a hard disk drive. In the discussion that follows, we will assume that the floppy drive is labelled **a:**, and the hard drive is designated **c:**. If your system has different drive designations, then substitute as necessary.

1.2.1 L-Edit Files

Your L-Edit disk contains the following files:

> ledit.exe
> ledit.tdb
> morbn20.ext
> morbn20.xst
> example.tdb
> xsect.tdb
> xsect.xst
> scna.spc

The main program is **ledit.exe**. When L-Edit is launched, it always looks for the file named **ledit.tdb** for setup information. This provides the technology base for your designs including the definitions of each layer and other important information. Both ledit.exe and ledit.tdb must be available when launching L-Edit.

A sample L-Edit layout is provided in **example.tdb**; this is shown in Figure 6.1 of Chapter 6. The file named **xsect.tdb** uses **xsect.xst** to provide an example of the Cross-Section Viewer; this drawing appears on the cover of this book, and is discussed in Section 8.5 in Chapter 8. A typical set of SPICE LEVEL 2 MOSFET parameters are provided in the text file **scna.spc** for use in circuit simulation.

In addition to the files listed above, six subdirectories have been included on the L-Edit disk. These are as follows:

> XSECT - Cross-Section Viewer data files;
> EXTRACT- Data files for Circuit Extractor;
> SAMPLES- Miscellaneous sample drawings;
> PRTSETUP- Alternate drawing background screens;
> TECH- CMOS technology files;
> MUMPS- Micromechanical technology files.

The XSECT, EXTRACT, and TECH provide the database for various CMOS processes, and are discussed in the text. The remaining files have been included to allow you to experiment with other applications of L-Edit. Note in particular the

---

MUMPS database for micro-mechanical submissions.

1.2.2 Running L-Edit from a Floppy Drive

Simply insert the disk into the disk drive, change to the floppy drive by typing

[Enter] means
to push the
"Enter" key.

> **a:**

followed by [Enter]. Assuming that your prompt is a:, you can launch L-Edit by first activating the mouse, and then following the directions below.

1.2.3 Using a Hard Disk Drive

Installing L-Edit onto a hard disk drive is recommended, as it provides much faster operation. To install L-Edit onto your hard drive, insert the L-Edit disk into drive a:. Then, make sure that you are in the root directory of the hard drive by typing

> a: **c:** [Enter]

Next, create a new subdirectory to hold the files. We will name the subdirectory LEDIT using the command

> c:**md ledit** [Enter]

After the subdirectory is created, copy all of the files from the floppy by typing

> c:**copy a:*.* c:\ledit** [Enter]

As a minimum, you must include ledit.exe and ledit.tdb; the remaining files and subdirectories are optional. The program can now be launched from your hard drive from the subdirectory \ledit. You may wish to modify your autoexec.bat file to include the \ledit subdirectory in a path command.

1.2.4 Launching L-Edit

L-Edit requires that a Microsoft-compatible mouse is installed for use as a drawing and pointing device. Follow the instructions provided with the mouse to install the device. Usually, a file with the name mouse.com activates the mouse by the command

> c:**mouse** [Enter]

Once the mouse is installed, change to the L-Edit subdirectory by means of

> c:**cd ledit** [Enter]

(unless you have included \ledit in a path command). L-Edit is launched by simply typing

> c:**ledit** [Enter]

This should result in the L-Edit screen information screen. Clicking a mouse button clears the information screen, and places you into the program.

1-4

Introduction to L-Edit

1.2.5 Loading a Data File while Launching

L-Edit can be also launched so that it automatically loads a design file. Suppose that you have a layout that has been saved under the name mylayout.tdb. Typing the command

> c:**ledit mylayout.tdb** [Enter]

will launch L-Edit and load the file named **mylayout.tdb** in the work buffer.

1.2.6 Using a RAM Disk

If the program is a little "sluggish" on your system, you might consider installing L-Edit on a RAM drive. This provides a significant improvement in the speed of some operations, particularly when using the more advanced features of circuit extraction and design rule checking. If you have 2MB or more of system memory (RAM) on your computer, you should consider this option. The software necessary for configuring a RAM disk is available in MS-DOS 5.X and higher using the VDISK option (or equivalent) in your config.sys file. Allocating about 1 MB of space on the disk is sufficient for the program and all of the data files in the root directory of the L-Edit disk. The size of the RAM disk can be reduced by copying only the necessary files. As mentioned before, you must have ledit.exe and ledit.tdb available to launch the program.

Let's assume that you have a RAM disk that is labelled as e:. The simplest way to launch L-Edit to run from the RAM disk is to create a batch file in c: that copies all of the files from the hard drive to the RAM disk, and then launches the program. An ASCII text editor may be used to write the batch file, or, you can create the file directly from the keyboard (console) using the following commands:

> c:**copy con: draw.bat** [Enter]
>
> **cd\ledit** [Enter]
>
> **copy *.* e:** [Enter]
>
> **e:** [Enter]
>
> **ledit** [Enter]
>
> **^Z** [Enter]

where **^Z** is obtained by pressing function key F6 on the keyboard. This sequence creates a batch file named **draw.bat**. To launch L-Edit, first check that the mouse is installed, and then type

> c:**draw**

The batch file will then boot from e:, and it becomes the default drive.

1.3 Exploring L-Edit

Launching L-Edit should result in the screen shown in Figure 1.1. The screen consists of several distinct areas as shown. This includes the menu bar, the drawing

---

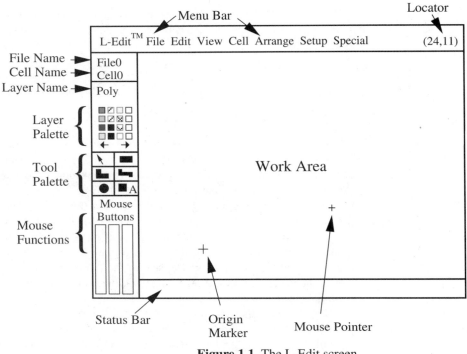

Figure 1.1. The L-Edit screen

area, the layer selections, and the mouse functions. Each of these is discussed below.

The Menu Bar

The Menu Bar provides access to most L-Edit commands.

The menu bar is the region at the top of the L-Edit screen. It provides access to all of the features of the layout editor including features such as disk operations, view control, and the design rule checker.

L-Edit operations are grouped into the categories **File**, **View**, and **Special**. To see the commands in each category, just use the mouse to point to a category, then push and hold the left mouse button. This drops down the window that lists the operations that are available. Hold the mouse button down and move the mouse until the desired operation is highlighted by the mouse pointer; releasing the button activates the chosen action.

Mouse Buttons

Mouse buttons are used to select operations.

The mouse is used as the primary drawing and control device. L-Edit provides a visual indicator of the mouse button functions in the lower left side of the L-Edit screen. This is convenient to remember, as the functions vary with the operation in progress. The example screen shows a 3-button mouse; if a 2-button mouse is used, L-Edit will display the functions in the appropriate format.

This book assumes that you have a 3-button mouse installed, and we will refer to the buttons by position: Left, Center, and Right. **If you are using a 2-button mouse, then only the Left and Right functions are displayed on the screen.** To access the Center button function, hold down the Alt- key while pressing the Left button; the menu will change and display the center function. Also try holding the Ctl (control) and Shift keys, as these often provide another command option.

Once the mouse has been used to point to the desired location, there are five main operations that we will define:

Press and hold. This means to depress a button and not to release it until the operation is completed.

Click. A click is a simple press-and-release of a button.

Double-click. A double-click is defined as being two clicks in rapid succession.

Move. This means to physically change the position of the mouse.

Drag. A drag operation is accomplished by combining a Press and Hold and a Move.

It is important to note the difference between a Move operation and a Drag operation. These operations allow you to draw and access the various commands in the L-Edit program.

Most L-Edit commands can also be accessed from the keyboard. The keystroke equivalent, if one exists, can be found listed in the command windows. These often use a combination of a key that is pressed simultaneously with the control key. In the book, the control key will be denoted by "^". For example, "^A" means push the "control" and "A" keys at the same time.

The Work Area

Most of the screen is for drawing, so let's start with a short description of the work environment. The **work area** is the region where the layouts are constructed, and is analogous to a large sheet of paper. The work area is that portion of the entire drawing that falls within the L-Edit screen. This is shown in Figure 1.1. The work area that appears in the L-Edit screen can be adjusted using the **Zoom Out** and **Zoom In** commands under the **View** heading; the "-" and "+" keys provide the same function from the keyboard. Device and circuit layouts are usually performed using the **Zoom In** to "magnify" the drawings, while larger sections of the chip can be viewed in a single screen by stepping the zoom out.

Locator

Points on the drawing area are defined by a pair of grid coordinates (x,y); the current position of the mouse pointer is shown in the upper right-hand corner within the menu bar. Under normal operation, all objects in L-Edit are referenced to this coordinate system. The origin (0,0) is marked with a cross-hair, but it may not be within the viewing area of the screen when L-Edit is booted. "Panning" refers to the ability to move the viewing area over different regions of the plot, and may be performed at any zoom level. Arrow keys may be used to pan in L-Edit.

File Name

The file name indicator lists the name of the current file that is loaded into L-Edit. When you save your work, the information will be stored on disk using the file name that is shown here.

Cell Name

Cells are used as building blocks in large designs.

A cell is a basic building block for a VLSI design. For example, a simple parallel adder can be made using full-adder cells that are connected in a particular way. A file can contain many cells. The cell name shows the name of the current block that you are editing. Cells can be stored and replicated within the layout, making them very useful for the design of large networks.

Layer Name

An integrated circuit is made up of various layers of materials, such as polysilicon, metal, etc. The layer name tells you the current layer that has been selected for drawing. The layers are determined by the setup file **ledit.tdb** that is loaded when L-Edit is launched. All layers can be accessed through the Layer Palette on the left side of the screen.

Layer Palette

Integrated circuit layout is based on the patterning of each material layer in the processing sequence. Every layer must be defined in order to design the circuit properly. Layout editors usually distinguish among the various layers by assigning a distinct color and/or pattern to each material. L-Edit provides a graphical menu of the available layers on the left side of the screen. To specify a layer, simply use the mouse to point and click at the appropriate box in the palette. All objects will assume the chosen color until the selection is changed. The name of the chosen layer will appear in the Layer Name line, and also in the status bar at the bottom of the screen.

Drawing Tools

L-Edit provides predefined tools for drawing polygons and other shapes. These are shown on the left-hand side of the screen. Drawing an object is very straightforward. Simply point to the desired shape, click to highlight it, then move the pointer to the desired position on the layout area. Pressing the mouse button and holding it down while moving the mouse (dragging the mouse) will draw the object; the size of the object is changed by moving the mouse. Releasing the button sets the final size of the object. Drawing operations are examined in more detail below.

The Status Bar

The status bar is used to provide information on the current operation. For example, the name of the selected layer will appear in the status bar when drawing or selecting an object. This acts as verification of the operation in progress.

1.3.1 Interacting with L-Edit

The mouse provides direct control of most program operations. It is used to draw and access the menus. Let's examine how to communicate with L-Edit; drawing is discussed in the next section.

The file structure of the L-Edit program provides access to commands using the Menu Bar. As discussed above, pointing to a command group name (such as Edit) and pressing the left button gives a drop-down menu; the menu remains open so long as the button is depressed. A command may be selected by dragging the pointer down until it is highlighted; releasing the button activates the line.

Certain types of operations will require you to provide L-Edit with input data. This is accomplished using a **dialog box** that appears in the Work Area as needed. When a dialog box is being displayed, all normal program operations are suspended. The box is used to request information, such as a filename or an option within the command. The keyboard is used for text input, while the mouse provides navigation within a box if needed.[1]

Chapter 7 provides a listing and short description of every L-Edit command. This has been provided in a reference format to help you work with all features of the program.

1.4 Drawing with L-Edit

Learning to work with a program is best accomplished with hands-on experience, so let's begin by drawing a simple rectangle.

This indicates an L-Edit example.

1. Point to a layer box and click it. Any box will do for this exercise.

2. Point to the rectangle in the Tool Palette area. Double-click the left-button to select this drawing tool.

3. Move the mouse pointer to the Work Area. Press and hold the left mouse button to define one corner of the rectangle; then, drag the mouse. This will result in the outline of a rectangle with dimensions that vary as you change the location of the mouse. The behavior of the size when the mouse is dragged is an example of "rubber-banding." If you release the button, the shape will be stored as it appears, and the object will take on the color of the selected layer.

As you can see, drawing in L-Edit is very simple! Note that we have highlighted the word "object" in Step 3. above. In the language of layout editors, every geometrical construct is referred to as an **object**, regardless of its exact shape. We will use the term **object** quite frequently in this book.

Now let's examine some of the important aspects of creating objects in L-Edit.

1.4.1 The Layout Area

All drawing is done on the Layout Area. This can be visualized as a large x-y plane as discussed above. Points on the Layout Area are defined by the coordinate pair (x,y). All geometrical objects created in the Layout Area have vertices defined by

---

[1] The Tab key may also be active. It is used to move from one data field to the next.

sets of coordinates. 30-bit signed integers are used for both x and y, so that the coordinates are in the range $\pm\, 2^{29} = \pm536,870,912$. The relationship between an internal unit and an absolute unit (such as a micron or centimeter) is defined by the setup file.

The Layout Area provides grid points that are useful for drawing objects of a particular size. The grid can be turned on or off using the appropriate command in the View grouping of the Menu Bar (Show Grid or Hide Grid). Since grid points are used primarily for detailed drawings with a close-in Zoom factor, they are not visible when viewing large regions of the Layout Area.

1.4.2 Basic Navigation

The view seen in the Work Area of the L-Edit screen is controlled by two groups of commands, **Pan** and **Zoom**.

Pan

Pan commands allow you to move the viewing area to the desired location. The basic movements are vertical and horizontal, and are controlled by the arrow keys as summarized in Figure 1.2. Note that pressing the **Shift** key with an arrow key translates the viewing area to the edge of the geometry. When using the Pan command, keep in mind that the objects contained within the Work Area move in a direction opposite to that shown by the arrow, since it is the view that is changing, not the location of the objects.

L-Edit has an Auto-Pan feature that allows you to draw objects that move beyond the viewing area. As the mouse is moved toward the viewing boundary, the program will automatically Pan to follow the mouse pointer.

Zoom

The zoom commands provide a method for obtaining "up-close" and "far-away" views of the drawing area. As mentioned above, these functions can be accessed from the **View** group of the menu bar. Alternately, keyboard commands can be

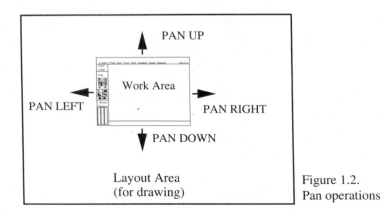

Figure 1.2.
Pan operations

(a) ZOOM OUT
 This allows you to see a larger layout area, but less of the details.

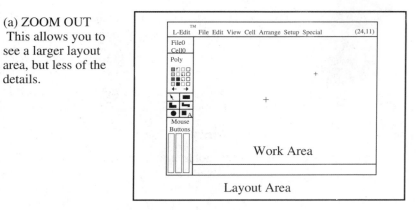

(b) ZOOM IN
 This allows you to see more of the details of the drawing over a small section.

Figure 1.3. The ZOOM operations

used: the "+" key is equivalent to Zoom In (for increased magnification), while the "-" key gives Zoom Out action. Figure 1.3 illustrates the main concept of a zoom function.

Two shortcuts can also be accessed from the keyboard. To see the entire drawing, press the **Home** key; L-Edit will automatically set the zoom to include all of the objects. If you would like to zoom out as far as possible, press the **End** key.

L-Edit also allows you to Zoom using the mouse. Pressing Z on the keyboard activates this feature. Point to the desired region, then use the buttons to perform Zoom Box (Zoom In on the region), Pan, or Zoom Out.

1.4.3 Drawing Tools

There are six types of drawing tools in the Tool Palette:

- **Arrow**: For pointing and selecting objects
- **Rectangle**: Draws rectangular objects
- **Polygon**: Allows you to construct general polygons

- **Wire**: Acts like a wiring tool with a predefined width
- **Circle**: Draws circular objects
- **Port**: The port tool is used to define signal entry points, or to apply text labels

The six regions of the tool palette are identified in Figure 1.4. These are shown in normal 90° mode, which is the most common for integrated circuit layout.[1]

Arrow Tool

The arrow tool is used to **select** objects or to define regions on your drawing; it does not draw a geometrical object. To use the arrow tool, select it from the tool palette by pointing and clicking, and then move the pointer to the work area. To select a single object, point to the object and click. Visual verification of object selection is a darkened outline. If the pointer is placed near a group of objects (i.e., not inside of a particular object) when the mouse is clicked, the nearest one will be selected.

To select a group of objects, use the Arrow Tool to draw a box that contains every object; objects on the border of the box will not be included. Releasing the mouse button highlights every object in the selected set.

Once an object, or set of objects, is selected using the Arrow Tool, the Edit commands can be used to perform operations such as move, reshape, duplicate, and others.

Rectangles

To draw a rectangle, just point to and click on the rectangle in the tool palette, and choose a layer from the Layer Palette. Moving the mouse pointer to the work area activates the drawing process. Depressing the left mouse button defines the **anchor point** for the rectangle; this is one corner of the object. Dragging the mouse (holding down the button as the mouse is moved) creates an outline of the rectangle. Releasing the mouse button completes the drawing process.

Polygons

The polygon tool is useful for creating regions that have turns and corners. Select the tool by pointing and clicking the left mouse button. Once in the work area, the left button defines the vertices, the center button is labelled Backup (used to undo the previous operation), and the right button labelled End finalizes the object. The

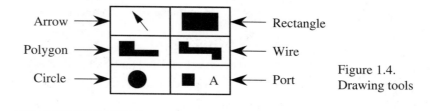

Figure 1.4.
Drawing tools

---

[1] Layouts that only allow 90° angles form a Manhattan geometry.

anchor point is the first corner, and is defined by the first click of the left button. Use the Move operation to draw. Each subsequent click on the left button creates another vertex, and the object is shaped by the Move mouse operation. (Note that the buttons should not be depressed during a Move operation.) When the object has the desired shape, push the right button to end the drawing process.

Wire Tool

The wire tool is used to draw lines with a specified width. The operation is similar to that used for polygons. Point to and click on the wire tool, then move to the drawing area. The first click with the left button defines the anchor point. Moving the mouse creates the line, and vertices are created by clicking the left button. Use the right (End) button to complete the object.

Figure 1.5 summarizes the three primary objects discussed above. The drawing was taken directly from L-Edit using a ZOOM factor that shows the grid points. Initially, you may find it easier to create layouts using only the rectangle tool. However, learning to use the polygon and wire tools will simplify the process.

Circles

Circular objects are easily constructed using the circle tool. Only the Left and Right buttons are used in the drawing process. The anchor point is the center of the circle, defined by the first click of the left button. Dragging the mouse changes the radius of the circle, and the size is finalized using the right button.

The circle tool is useful in certain applications such as designing micromachine geometries. Most CMOS fabrication lines, however, cannot produce circular shapes on the masks.

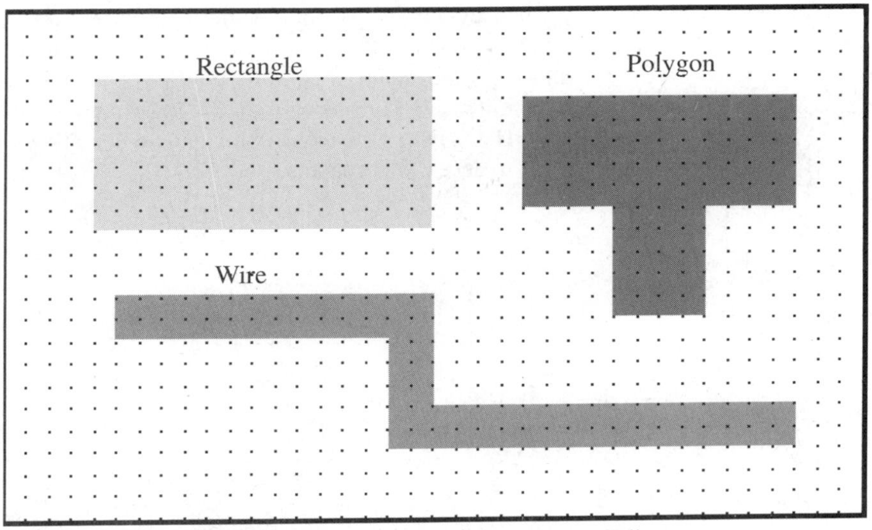

Figure 1.5. Examples of objects in L-Edit

Ports

A port on the layout drawing corresponds to a signal entry or exit point; it is defined relative to a drawn object, such as a rectangle or a line. To define a port, move the mouse to the object and click the left button. A dialog box will appear, prompting you to type in a name for the port; all ports must have names in L-Edit. This action results in the port name appearing on the layout. Ports are convenient for labelling lines, such as power supply and ground connections, or for adding text labels to your drawing.

Non-Manhattan Geometries

L-Edit also allows you to construct a drawing with non-orthogonal polygons. To access this mode, double-click on the polygon or wire box. The Tool Palette will change icons. The first set that appears are for 45° angles; the palette for this case is shown in Figure 1.6(a). If the tool is clicked another time, L-Edit enters the general polygon drawing mode, which displays the palette shown in Figure 1.6(b). A third click returns you to the Manhattan (90°) geometry mode. Caution must be exercised if non-orthogonal polygons are used in an integrated circuit layout, since the fabrication process may not allow for these types of shapes. In the discussion presented here, we will confine ourselves to right angle geometries as being the most useful. However, in high-density design, you may want to access these drawing functions.

1.4.4 The Layer Palette

Integrated circuits consist of several material layers, each patterned in a specific way to form a three-dimensional structure. The Layer Palette allows you to choose the desired layer as you design the circuit. Each entry corresponds to a predefined material layer in the integrated circuit fabrication process. Selection is accomplished by simply pointing to the appropriate box and clicking the mouse. Each layer is coded with a distinct color and/or pattern, and the name of the layer is shown in the Status Bar at the bottom of the screen as the mouse pointer passes over it. Once a layer is chosen, the layer name is also displayed in the same location. Using the arrow icons in the palette box allows you to scroll the palette left or right to access other layers.

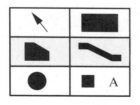   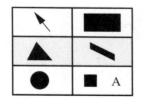

    (a) 45° mode   (b) All angle mode

Figure 1.6. Non-orthogonal drawing tools

The Layer Palette window displays 16 layers at a time in a 4×4 grid arrangement. L-Edit allows more layers to be defined using 4-entry columns. Layers outside of the palette window can be accessed by using the left and right arrow icons as shown in Figure 1.7. The number of available layers depends on the technology file that is currently loaded into L-Edit, and any **derived** layers that have been created. Derived layers are new layers that are defined using logical operations among basic technology layers; these are discussed in more detail in Chapter 3 and Chapter 9.

1.5 Fundamentals of Object Editing

L-Edit has a simple user interface to make object drawing intuitive and easy. Once the objects have been created, the program provides for several editing operations that can be used to modify the properties of objects.

In this section, we will examine several of the basic object editing operations that are possible with L-Edit. More advanced features are discussed in later chapters in the context of CMOS chip design and layout. Also, Chapter 7 provides a listing of all editing commands and a brief description of their usage.

1.5.1 Object Selection

To manipulate an object, you must first specify which object (or group of objects) are to be modified. This is accomplished by a Select operation.

Select Operations

A single object is selected by the point and click operation. The Mouse Button indicators always tell you the functions that are permitted in the current mode. If you point to the interior of an object and click, it will be selected. Or, if the pointer is outside of a group of objects, the nearest one will be selected. The selected object is identified by having a darker outline than the unselected objects. A group of objects is selected by using the Arrow Tool to drag a box around all of the objects. Every object must be contained within the box to be included with the group; those on the boundary will not be selected.

Once a group of objects has been selected, you may add to the group by pressing Shift on the keyboard and dragging a box around the new objects. This operation is called **Extend Select**, and allows you to preserve an existing set while adding new objects to it.

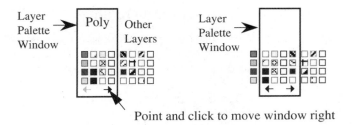

Point and click to move window right

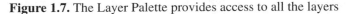

Figure 1.7. The Layer Palette provides access to all the layers

---

The **Unselect** operation is used to remove objects from the group. To effect an **Unselect**, press the Alt key while dragging a box around the objects that you wish to remove from the group; note that the right mouse button specifies the unselect operation.

Using the keyboard command ^A selects all objects in the layout. This may also be accomplished using the command in the **Edit** window of the Menu Bar. The operation **Unselect All** can be performed using either Alt-A from the keyboard, or the **Edit** Menu entry.

Layer Hiding

Layers can be hidden using the **Hide** command. Hidden layers are not shown on the screen, and are not included in the select process. To hide a layer, point to the layer in the Technology Palette and click the **Hide** button; all objects on the hidden layer are not shown on the screen, but the information remains stored in the edit buffer. If a **Select** operation is performed when a layer is hidden, the objects will not be contained in the selected group after the layer is made visible again.

1.5.2 Moving Objects

The location of an object may be changed by depressing the Move/Edit button while dragging the mouse pointer to the new location. This places the object at a new set of coordinates.

To move a single object, place the pointer inside of the object, click the Move/Edit button, and drag the object to the desired location. If you wish to move a group of objects, then you must first perform the Extend Select operation.

1.5.3 Changing the Size and Shape

Once an object is drawn, it can be resized or reshaped using the edit operations. Simply place the mouse pointer on an edge or a corner, and use the Move/Edit button while dragging the mouse. Releasing the button sets the new size or shape.

If you wish to add a new vertex to a polygon or a wire, select the tool and then switch it into the All-angle mode. Next, position the mouse pointer over the edge of the object where the vertex is to be added. The function is enabled using the Add/Edit function, which is the Move/Edit button combined with the Control key. Dragging the mouse shows the object with the new vertex. Releasing the Add/Edit button finalizes the shape of the object.

1.5.4 Cutting Objects

An object (or group of objects) may be cut from the diagram and removed using this command in the Edit window. Just select the object(s) and choose **Cut** from the Edit window, or ^X from the keyboard.

1.5.5 Copying Objects

An object (or group of objects) can be copied using the **Copy** command. You must be in the Draw mode (with a tool chosen) to activate this option. Select the object and then invoke the copy function using Control and the Draw button on the mouse; or, type ^C from the keyboard.

1.5.6 Arrange Commands

The Arrange window in the Menu Bar provides several useful commands that change the orientation of an object.

Rotate

This command rotates the selected object by 90 degrees in a counterclockwise sense about its center point. If this is applied to a group of objects, the rotation is about the center of the group.

Flip Horizontal/Flip Vertical

The **Flip** commands perform a flip of the selected object about an axis that passes through the center.

Cut Vertical/Cut Horizontal

A **Cut** command divides an object through an axis that is selected by the mouse pointer.

Merge Selections

This command merges intersecting objects on the same layer into a single object. Just select a group of objects and execute the command. All objects on the same layer that intersect will be merged into single objects.

1.6 Learning L-Edit

Although L-Edit is a very powerful layout editor, it is quite easy to learn. There are two levels of familiarity:

- Learning the basic operations, and
- Applying the program to a specific task.

Basic operations are best learned by trial and error. Practice by drawing objects using the tools. Start with the rectangle tool, then try to draw wires and polygons. Change the layers to get a feel for how layer selection is accomplished. Once objects have been drawn, move and edit them until you feel comfortable with the mouse operations. Don't worry about making mistakes; they can always be corrected. Or, if all else fails, just reboot your computer! You will discover that the L-Edit environment provides a friendly user interface.

Applying L-Edit to a particular task, such as designing an integrated circuit or a micro-machine, requires that you understand why a layout editor is needed to begin with. Once the usage is defined, then the specific details can be worked out. In the next five chapters of this book, we will examine the use of a layout editor in designing digital CMOS chips. However, L-Edit can be used in any application that requires the creation and manipulation of geometrical objects.

1.7 Monitoring Memory Usage

As was mentioned previously, this version of L-Edit cannot access RAM above the 640K DOS limit. While this should not cause a problem for any classroom-type designs, a complex layout or a large cell library[1] may start to use up the available memory. Should this situation arise, L-Edit will warn you about the problem. Do not ignore the warnings! If L-Edit runs out of memory, you may lose some or all of your work.

It is recommended that you monitor memory usage while working, especially if you are creating a large layout. First, choose Status from the L-Edit window of the Menu Bar:

The Status command allows you to monitor memory usage.

This action results in the Status window shown in Figure 1.8. The most important number to monitor is the "Free Memory" statement on the last line. This is given in both number of bytes and percentage (%) values. Be sure to check the amount of free memory as you draw.

1.8 Saving Your Work

It is very important that you save your drawing at regular time intervals. To perform this operation for the first time, point and click on **File** on the Menu Bar, depress the

Memory Status	Count	Bytes	%
Memory Avail:		194720	100%
Files:	2	7612	4%
Cells:	4	168	0%
Instances:	0	0	0%
Boxes:	61	244	0%
Circles:	0	0	0%
Ports:	3	24	0%
Wires:	0	0	0%
Polygons:	3	100	0%
Undo List:		0	0%
Select List:		0	0%
Cut/Paste Buffer:		0	0%
Misc:		0	0%
Total used:		8156	4%
Free Memory:		186564	96%

● All Files ○ This File ○ This Cell

[OK]

Figure 1.8.
The L-Edit status box

___
[1] Cell libraries are discussed in Chapter 6.

mouse button while moving the mouse until **Save As...** is highlighted as shown in Figure 1.9, then release the button. A dialog box will appear asking you to provide a file name. It is recommended that you choose a distinctive name that will help you remember the contents; the default names provided by L-Edit are file0.tdb, file1.tdb, etc., where the number is automatically incremented to avoid duplicate names. After the file has been named, the file can be saved using the **Save** command in the **File** window.

If you have created a large layout drawing, the **Save** operation becomes crucial for two reasons. First, the obvious: you do not want to lose your work if a power outage (or some other catastrophe) should occur[1]. The second reason has to do with the RAM limitation of this version of L-Edit. If the amount of available RAM gets too small, L-Edit may not have sufficient memory to store the next operation, and you may lose some or all of the information in the file.

1.9 Advanced Features

The Student Version of L-Edit includes several advanced functions that place you into a realistic CMOS design environment. These are contained in the **Special** entry of the menu bar, and are accessed in the standard way.

Design Rule Checker

Design rules are a set of minimum size and minimum spacing values that should be obeyed to ensure maximum yield in the chip fabrication sequence. High-density

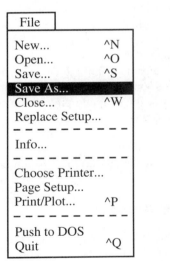

Figure 1.9. The Save commands

---

[1] In the classroom environment, problems like this often arise during the last week of classes, just before your design project is due.

logic design is accomplished by close-packing of the devices and interconnects, so that the design rules may be unintentionally violated. The design rule checker (DRC) included with L-Edit will analyze your layout, determine if any design rules have been violated, and report on violations that have been found. Violations can be specified graphically or in a text file format.

Circuit Extractor

A layout drawing is a physical representation of a circuit made up of elements such as resistors, capacitors, and transistors, that are wired together. Although a schematic diagram is used to construct the layout, the physical implementation always introduces parasitic elements that affect the performance. Also, it is important to ensure that the layout does not contain any errors.

L-Edit provides a circuit extraction algorithm that translates your layout into a SPICE-compatible text file that can be used to simulate the network. Elements such as MOSFETs are described by their node connections and geometries using a set of predefined rules for the technology. This allows for direct verification of the layout. Moreover, the circuit extractor can be programmed to include parasitic resistance and capacitance in the devices and interconnects, so that the resulting simulation is directly based on the layout geometry.

Circuit extraction is very important for learning the interplay between the layout and the resulting electrical performance, i.e., chip design.

Cross-Section Viewer

This unique feature of L-Edit allows you to see the "side view" of your layout design. Cross-sectional views for devices such as MOSFET are drawn according to the "top view" patterns that are drawn in the work area. The viewer lets you watch each layer being grown and patterned, and clearly illustrates how the layers stack to form a 3-dimensional integrated circuit structure. Moreover, the drawing sequence is correlated to the steps used in the fabrication process. Cross-sectional drawings of this type are useful for understanding the relationship between the layout and the resulting devices. Parasitic coupling problems between interconnect lines can also be spotted using the drawings.

The features described above are introduced in the book using explicit examples in CMOS circuit design. They are also documented in the later chapters of the book.

1.10 Plan of the Book

This book is designed to be used as both a tutorial in CMOS chip design and a basic reference manual for the L-Edit program. You may find it useful to work through the tutorial chapters even if you have already studied CMOS circuits. If you are an experienced user of layout tools, then you may find that the information in this chapter is sufficient to learn the main features of the program.

Chapters 2 through 6 provide an introduction to the basic concepts in the design and layout of digital CMOS integrated circuits. Included are individual chapters on fabrication, physical design and layout, MOSFET, logic circuits, and systems design. We have attempted to present most of the basic concepts that form the foundations of the subject.

The tutorial approaches CMOS design in both general and specific terms. First, the theory is examined, and important equations and graphs are introduced as needed. Specific reference to L-Edit is made at selected points in the discussion. These give step-by-step instructions on using the L-Edit program to layout circuits to illustrate the topic at hand. Additional practice exercises have also been included. Examples using L-Edit are marked with the icon shown below; this allows you to quickly locate instructions as needed.

Look for this
icon to find
L-Edit examples.

Although the tutorial has been written as a short introduction to CMOS chip design, it is far from a complete treatment. To improve your expertise in the field, you should combine the material presented here with your course studies. To become proficient in chip design, it is very important to learn how the circuit operation is related to the layout and fabrication at the silicon level. Take the time to study each layout from both the electrical and physical viewpoint.

The latter portion of the book (Chapters 7 through 11) forms an abbreviated reference manual for L-Edit operations and the advanced features, such as design rule checking and circuit extraction. No general tutorial comments are given in these chapters. Sufficient documentation has been included to allow you to use and customize L-Edit as desired.

The authoritative guide to using the L-Edit program is the *L-Edit Manual* written by Tanner Research. It provides detailed, step-by-step instructions for every feature of the program. The Manual is provided with the full Professional Version of L-Edit, or it may be purchased directly from Tanner Research for use with the Student Version.

Please note that, due to the low cost of this software package, it is not possible for Tanner Research or the author to provide any support for the Student Version of L-Edit. However, Tanner does offer full customer support with the full professional version, including individualized help and updates.

1.11 A Note About Copy Protection

The floppy disk containing the Student Version of L-Edit has been provided without any copy protection to make it easy for you to use. You are permitted to make one (1) copy of the disk for backup purposes only. However, the license does not allow you to install and/or run the program on more than one machine at a time.

We would like to ask you to observe these simple rules. In particular, please do not make copies of the disk for others to use. We have strived to keep the price of the software/book combination low to allow students the opportunity to use a professional VLSI CAD tool on their own computer. This is only possible if everyone respects the copyright.

1.12 Mumps

The Student Version of L-Edit includes a set of files in the MUMPS subdirectory. MUMPS is an acronym for *Multi-User MEMS Process*, and MEMS is an acronym for *MicroElectroMechanical Systems*. In everyday terms, this field is sometimes referred to as *micromechanics* or *micromachines*. MUMPS is an ARPA-sponsored activity that provides for low-cost access to a standard fabrication process that is run by the Microelectronics Center of North Carolina (MCNC). This is similar to the MOSIS arrangement for CMOS integrated circuits.

The files include a technology base for designing MEMS in L-Edit obtained as a last-minute addition to the software package. As such, the book does not include any specific references to the files. The file mumps.tdb provides the basic technology base that can be used for design. For more information on MUMPS, please contact MCNC directly.

Chapter 2

CMOS Fabrication

In this chapter, we will review the basic processing sequence used to fabricate CMOS integrated circuits. This will provide a background for the layout and physical design of the chip.

2.1 Overview of Integrated Circuits

There are several approaches that can be used to describe an integrated circuit (IC). In the basic sense, an IC is an electronic network that has been fabricated on a single piece of a semiconductor material such as silicon. The silicon surface is subjected to various processing steps in which impurities and other material layers are added with specific geometrical patterns. The steps are sequenced to form three-dimensional regions that act as transistors for use in switching and amplification. Passive elements, such as resistors and capacitors, are not always included as elements in the circuit, but arise as parasitic elements due to the electrical properties of the materials. The "wiring" among the devices is achieved using **interconnect**s, which are patterned layers of low-resistance materials such as aluminum. The resulting structure is equivalent to creating a conventional electronic circuit using discrete components and copper wires.

In our examination of integrated circuit layout and design, we will define an integrated circuit as a set of patterned layers. Each layer has specific electrical characteristics, such as sheet resistance, and is patterned according to the layers above and below. Stacking different material patterns results in geometrical objects that function electrically as devices or interconnects.

Physical design transforms a circuit schematic into a silicon chip.

A layout editor such as L-Edit is used to design the patterns on each layer and accomplish the **physical design** of the chip. The drawings represent the patterning of each layer, and the overall image can be interpreted as the top view of the chip. Each layer is distinguished by a separate color on the computer monitor so that three-dimensional structures such as transistors can be designed. Although this level of abstraction is quite different from analyzing a circuit using a schematic diagram, it is easy to get used to. Physical design is discussed in detail in the next chapter.

2.2 Lithography and Pattern Transfer

Lithography is the process used to transfer the desired pattern to each layer of material on the chip. There are several steps used in the lithographic sequence. The major ones are:

- Drawing the patterns using a layout editor
- Preparing each pattern for physical transfer to the wafer
- Transferring the pattern on the wafer (called printing)
- Using processing techniques to physically pattern each layer.

We note that although a layout editor allows us to view all of the layers simultaneously, the lithographic sequence must be applied separately for each layer that makes up the chip. Using the completed design file, information is extracted for each layer in the process. This data is used to create individual **masks** for each layer; a mask is a plate of glass on which the pattern has been duplicated on a thin layer of chromium on the surface of the glass. The mask is transparent to light except in regions where the chromium acts to block light transmission.

Figure 2.1 illustrates a typical lithographic sequence where a doped region is patterned into the silicon substrate. The starting point is shown in Figure 2.1(a); at this point, an oxide layer has been grown on the surface of the silicon. Next, the wafer is spin-coated with liquid **photoresist** (sometimes referred to as simply "resist") and then dehydrated in an oven for a short time to yield a firm surface; this results in the coating drawn in Figure 2.1(b). Photoresist is a light-sensitive, organic polymer, with properties similar to ordinary photographic film. The image on the mask is optically projected to the surface of the photoresist during the exposure step schematically illustrated by the simplified drawing in Figure 2.1(c)[1]. Next, the resist is developed using appropriate chemicals. There are two types of photoresist, negative and positive. After exposure and development, negative photoresist hardens in regions that have been exposed to light, while areas that have not absorbed photons are soluble; positive photoresist is exactly opposite. The situation shown in drawing (d) assumes that a negative resist has been used; however, in fine-line patterning, positive photoresist is required.

The patterned photoresist layer is used as a mask for etching a material layer. In the present example, the original silicon dioxide layer will be patterned using a process such as RIE (**reactive-ion etching**). This gives the patterning shown in (d). Finally, the resist is removed, and the final structure is shown in (e). The patterned silicon dioxide may now be used as a mask for doping. In Figure 2.1(f), an ion implanter beam is directed toward the surface. Exposed silicon areas are doped by

---

[1] The mask is shown close to the surface for ease of visualization. In modern IC lithographic equipment, the mask information is contained on a **reticle**. The image on the reticle is projected onto the wafer using an optical printing system. Individual die are exposed using a **step-and-repeat** process across the wafer.

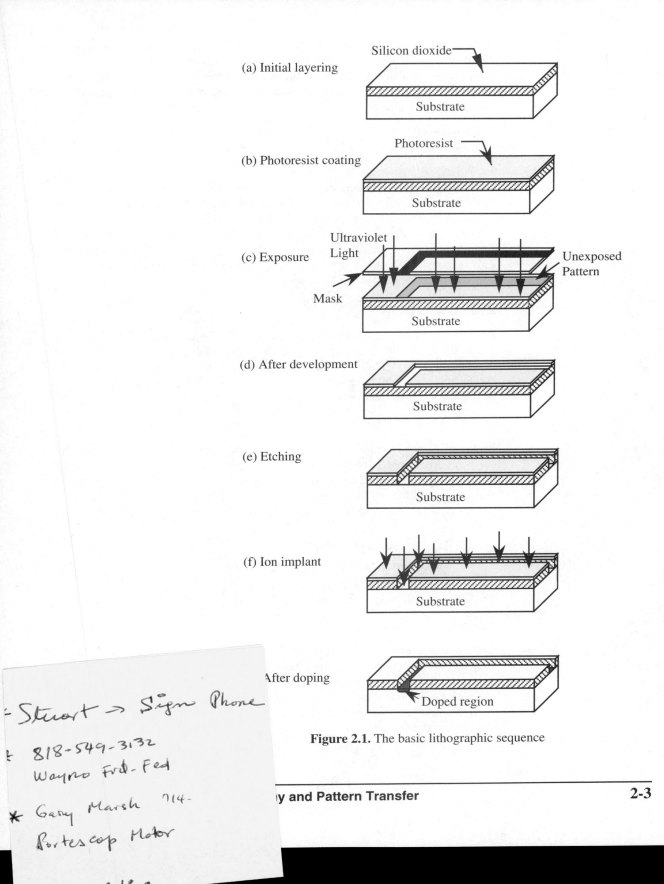

(a) Initial layering

Silicon dioxide

Substrate

(b) Photoresist coating

Photoresist

Substrate

(c) Exposure

Ultraviolet Light

Unexposed Pattern

Mask

Substrate

(d) After development

Substrate

(e) Etching

Substrate

(f) Ion implant

Substrate

After doping

Doped region

Figure 2.1. The basic lithographic sequence

the beam, while the oxide prevents the implant from reaching the covered regions. This results in the doped patterned region shown in Figure 2.1(g).

In fabricating an integrated circuit, the lithographic sequence is repeated for each material layer used to construct the device. The sequence is always the same:

1. Photoresist application;
2. Printing (exposure);
3. Development;
4. Etching.

The limitations of the patterning process, such as the minimum resolution and the minimum spacings, give rise to a set of mask design guidelines called **design rules**. These are discussed in more detail in the next chapter.

The previous example illustrated how doped regions in silicon are created. Of equal importance is the ability to pattern material layers above the silicon surface. This is shown on the next page in Figure 2.2 for a polysilicon line. It is important to note that the patterning sequence is unchanged.

2.3 CMOS

CMOS is an acronym for complementary metal-oxide-semiconductor. In everyday usage, the term CMOS can imply the technology, circuits, chips, or virtually anything having to do with the technology itself. CMOS has become the dominant silicon technology for high-density logic circuits, primarily because the transistors can be made very small, allowing for VLSI (very-large-scale integration) designs.

2.3.1 MOSFETs

MOSFETs (metal-oxide-semiconductor field-effect transistors) are the transistors used in CMOS integrated circuits. There are two types of MOSFETs: n-channel transistors that use negatively-charged electrons for current, and p-channel devices that conduct current by means of positively-charged holes.

MOSFETs have become the primary switching devices in high-density IC design because

- They are extremely small;
- The "drain" and "source" terminals are interchangeable;
- The device structures are very simple.

These points will be stressed throughout our discussion.

Structure

The basic features of an n-channel MOSFET are shown in Figure 2.3. The MOSFET terminals are called the **gate**, the **drain**, and the **source**. Also note that the connection to the p-type substrate gives the **bulk** electrode. The bulk is connected to the lowest voltage in the circuit, usually ground.

To represent an n-channel MOSFET in a layout editor, we note the following distinct material layers exist:

- p-substrate
- n$^+$ drain and source
- gate oxide (silicon dioxide, SiO$_2$)

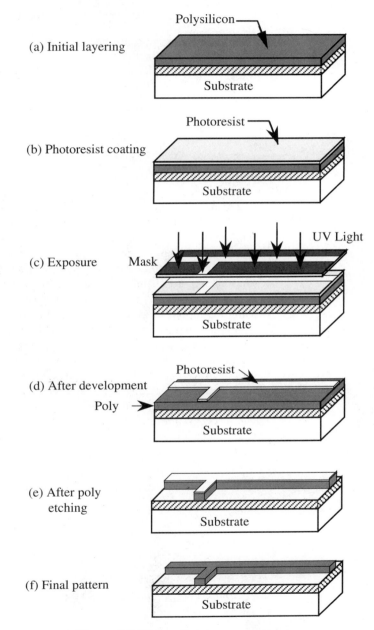

(a) Initial layering

Polysilicon

Substrate

(b) Photoresist coating

Photoresist

Substrate

(c) Exposure

UV Light

Mask

Substrate

(d) After development

Photoresist

Poly

Substrate

(e) After poly etching

Substrate

(f) Final pattern

Substrate

Figure 2.2. Poly patterning sequence

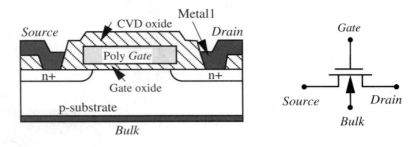

(a) Cross-section   (b) Circuit symbol

Figure 2.3. An n-channel MOSFET.

- polycrystal silicon ("poly")
- CVD oxide
- metal1

In general, oxide layers are implied in our drawings, as is the substrate. This leaves only the n^+, poly, and metal1 patterns to deal with. The presence of oxide is acknowledged by including a contact cut, which allows metal1 to be electrically connected to the n+ layers.

The primary geometrical parameters of a MOSFET are the **channel width** W (the extent of the channel as measured into the page), and the **channel length** L (the distance between n^+ regions under the gate oxide). These are more easily seen using the top view drawing shown in Figure 2.4. Note that the MOSFET channel length L is slightly smaller than the **drawn** channel length L' of the poly gate due to lateral doping effects. Current flow depends on the aspect ratio (W/L), and circuit design usually hinges on determining the aspect ratios for every transistor. We will examine this parameter in greater detail later in the discussion.

A p-channel MOSFET has the same geometrical structure as an n-channel device, but with reversed polarities, i.e., n-type regions are changed to p-type, and vice-versa. While this is straightforward to visualize, an added complication in a CMOS technology is that we need to construct both nFETs and pFETs in a common

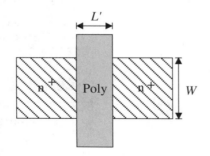

Figure 2.4.
Top-view of an
n-channel MOSFET.

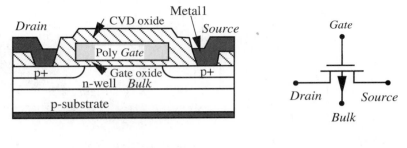

(a) Cross-section   (b) Circuit symbol

Figure 2.5. A p-channel MOSFET.

substrate. If we choose a p-type substrate, then we must provide n-type regions for use as the bulk of p-channel MOSFETs. This is accomplished by creating **n-well** sections in the substrate, resulting in the pFET structure shown in Figure 2.5. The layering scheme for this transistor is

- p-substrate
- n-well
- p^+ drain and source
- gate oxide
- polysilicon
- CVD oxide
- metal1

The primary layers that require patterning are the n-well, the p^+ regions, the polysilicon, and metal1. Note that the polysilicon and metal layers are common in nFETs and pFETs, and are therefore used as interconnects in circuit layout. As with the n-channel MOSFET, the primary geometrical parameters are the channel length L and the channel width W, such that the current is proportional to the aspect ratio (W/L). Figure 2.6 shows the top view of a pFET. It should be noted that the n-well is the bulk terminal for the pFET, and is usually biased to the power supply V_{DD}.

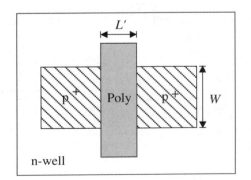

Figure 2.6.
Top-view of a
p-channel MOSFET.

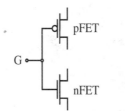

Figure 2.7.
A complementary pair.

2.3.2 Complementary Pairs

The *C* in CMOS means complementary. This terminology arises from the use of opposite-polarity transistor pairs in circuits. A **complementary pair** consists of an nFET and a pFET with their gates connected as illustrated in Figure 2.7. Note that we have introduced simplified circuit symbols for the transistors that do not explicitly show the bulk electrodes. Instead, the distinction between nFETs and pFETs is made by the absence or presence of an inversion bubble at the gate. The electrical properties of complementary pairs will be examined in detail in Chapter 5. For now, we will simply mention the fact that a CMOS integrated circuit requires that both n-channel and p-channel MOSFETs be fabricated in the same substrate wafer.

2.4 The CMOS Fabrication Sequence

Integrated circuit layout drawings represent the surface patterns of the layers that make up the chip. Both the geometrical and electrical characteristics of the structure rely on the details of the processing sequence. In this section we will examine the basic sequence for a typical n-well CMOS process.

0. **Start**. The starting point in our process is a p-type wafer. Electrically, this will serve as the bulk region for n-channel MOSFETs.

1. **Epitaxial Growth**. A thin layer of p-type epitaxial silicon is grown on the wafer, resulting in the view shown in Figure 2.8. The epi layer is used as the base layer for building devices. In the remaining cross-sectional drawings, we will only show the p-epi layer and will omit the substrate for simplicity. It should also be noted that the drawings are not to scale, but have been exaggerated for clarity.

2. **n-Well Formation.** In CMOS, the p-channel MOSFETs must reside in an n-type

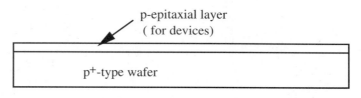

Figure 2.8. Epitaxial layer growth.

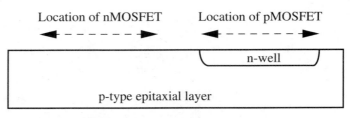

Figure 2.9. MOSFET placement.

background. These are obtained by adding an n-well as shown in Figure 2.9 to accommodate the pFETs. Electrically, the n-well region must be kept at the highest circuit voltage to ensure proper operation. This is usually the power supply voltage, denoted as V_{DD} in CMOS circuits. It is important to remember this point, since the power supply contact must be provided in the masking to ensure proper circuit operation.

3. Active Area Definition and Isolation. Since the transistor density in a VLSI design can be quite high, we need to be concerned about the problem of electrically isolating each device from its neighbors. Properly isolated transistors do not exhibit any direct conduction paths to any other devices, except for those that have been explicitly included through other layers. In this step of the fabrication sequence, the location of every transistor is defined; isolation is achieved by the next oxide growth.

First, we need to introduce some terminology. Consider the total surface area of the chip. The area is divided into two classifications, according to usage. An **active area** is a planar section of the surface where transistors (MOSFETs) are placed. **Field regions** surround the active areas, and constitute the majority of the surface area. Interconnect lines, such as polysilicon or metal patterns, are routed over the field, allowing us to "wire" the transistors together as required by the circuit.

Active areas are defined by patterned layers of silicon nitride (SiN_3), which is deposited onto a thin oxide layer on the silicon surface; the SiO_2 layer is known as the stress-relief oxide, and is used as a mechanical buffer between the nitride and silicon. A cross-sectional view of the wafer at this point is shown in Figure 2.10(a). We also note that a p^+ **field implant** is used to enhance the isolation[1]. (The field implant will not be explicitly shown in the remaining drawings.)

The silicon nitride layer is used in the next step known as the local oxidation of silicon (**LOCOS**). Exposing the surface of the wafer to a flow of oxygen-rich gas induces the growth of silicon dioxide in the field regions where the silicon surface is exposed, i.e., the field regions. In contrast, areas that are covered by nitride do not undergo the oxidation process, and silicon nitride itself does not oxidize. Since the growth of silicon dioxide requires silicon atoms from the substrate, the resulting **field oxide** (or, FOX) is recessed into the surface as shown in Figure 2.10(b). It is

---

[1] The implant increases the threshold voltage of a field region so that the voltage applied to an interconnect line will not create an unwanted conduction channel.

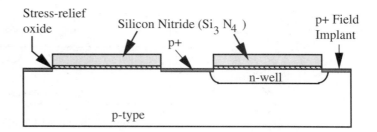

(a) Active Area Definition

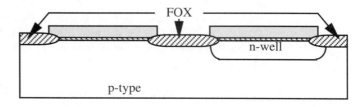

(b) FOX Growth

Figure 2.10. Active area definition and field oxide growth.

straightforward to show that, for a field oxide with a thickness of X_{FOX}, the field oxide is recessed a distance

$$X_{Si} \approx 0.46 X_{FOX}$$

(2. 1)

below the location of the silicon surface in an active area. Since silicon dioxide is an excellent electrical insulator, the recessed oxide region blocks lateral current flow (that parallel to the surface) between devices, achieving the desired isolation. In the literature, the terms field oxide and **recessed oxide** (ROX) are used interchangeably.

4. Gate Oxide Growth. After the field oxide growth is completed, the nitride and stress-relief oxide are removed. The critical gate oxide is grown next, giving the value of x_{ox}. This sequence is shown in Figure 2.11(a)-(b).

5. Polysilicon Deposition and Patterning. A layer of polysilicon ("poly") is deposited over the entire surface and then patterned by a lithography sequence. All of the MOSFET gates are defined by this step. Figure 2.11(c) illustrates the chip after this step. The polysilicon may be doped *in situ* (while it is being deposited). This lowers the parasitic resistance, which is an important consideration for high-speed integrated circuits. Other variations include adding a high-temperature (refractory) metal, such as platinum or titanium, to the polysilicon to create a **silicide** or **polycide** layer with small resistivity, or to coat the poly with a refractory

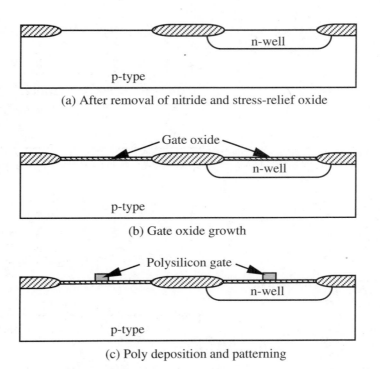

(a) After removal of nitride and stress-relief oxide

(b) Gate oxide growth

(c) Poly deposition and patterning

Figure 2.11. Gate oxide growth and poly gate formation.

metal layer.

6. pFET Formation. The formation of p-channel MOSFETs is accomplished during the next step. Photoresist is applied and patterned so that it covers all but the locations of the p^+ regions. Applying an ion beam consisting of boron ions to the wafer surface creates the p^+ drain and source regions of the pFETs, as shown in Figure 2.12(a). Accurate placement of these regions (relative to the gate) is accomplished by using a self-aligned approach where the polysilicon gate serves as a mask to the underlying channel area. Note that the polysilicon gate of the pMOSFET gains p-type acceptor doping. If *in situ* n-type dopants were added during the polysilicon deposition, then the final polarity of the gate (n-type or p-type) depends on which dopant species is dominant.

7. nFET Formation. Another lithographic sequence is used to define all nFETs. Donors (such as arsenic) are ion-implanted to dope the n^+ drain and source regions in the self-aligned transistors; this is indicated in Figure 2.12(b). After the implants are completed, a thermal annealing yields the cross-section shown in Figure 2.12(c). Both the n-channel MOSFET and the p-channel MOSFET structures can be identified after this step. Note that the nMOSFET gates are doped n-type in this step.

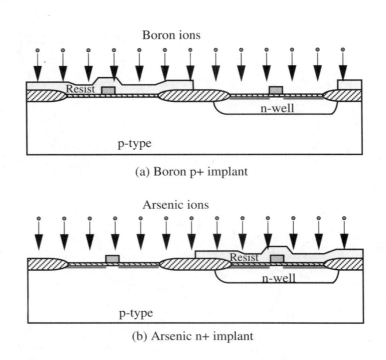

(a) Boron p+ implant

(b) Arsenic n+ implant

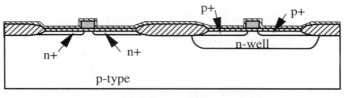

(c) After annealing

Figure 2.12. Drain/source ion implants.

8. Metal1. The entire wafer is coated with CVD[1] oxide. Contact cuts are patterned to provide metal1-active and metal1-poly electrical connections. A Metal1 layer is applied and patterned, resulting in the structure shown in Figure 2.13.

9. LTO Oxide and Via Definitions. Another CVD LTO layer of oxide is added, and openings (Vias) are created to provide for electrical contact to the next layer (Metal2) with underlying interconnect levels. The wafer at this point is shown in Figure 2.14(a).

---

[1] CVD stands for controlled (or chemical) vapor deposition. This type of oxide is formed by a chemical reaction, and does not require silicon from the wafer. It is common to use an LTO (low-temperature oxide) deposition for this and subsequent oxide layers so that underlying doped regions do not undergo any diffusive spreading.

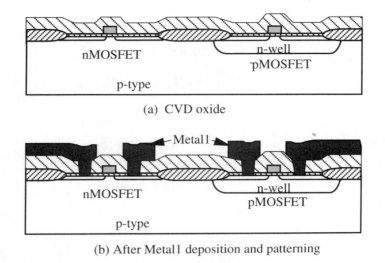

(a) CVD oxide

(b) After Metal1 deposition and patterning

Figure 2.13. Metal 1 layering.

10. Metal2 Deposition and Patterning. Metal2 is deposited and patterned, leaving the cross-section shown in Figure 2.14(b). This shows how the oxide cuts described in Step 9 allow the Metal2 material to contact the Metal1 layer by means of a **via**.

11. Overglass and Pad Openings. The final step is to add a protective layer over the surface. Typically, this consists of another layer of silicon dioxide followed by a layer of silicon nitride; the nitride is used because it acts as a good diffusion barrier against contaminants. Contact cuts in the overglass layer are required to electrically access the input/output pads on the final metal layer.

The sequence above describes all of the basic steps in an n-well CMOS process. However, state-of-the-art fabrication lines include many steps that are added to improve the device characteristics or improve the yield (the percentage of functional chips). As an example, **planarization** is a technique to "smooth" the surface prior to the deposition of the next layer; this makes the terrain less rugged, and provides for a more uniform layering surface in an effort to avoid breaks in the lines formed in the next layer. Lightly-doped drain (LDD) MOSFETs use two ion implant steps, resulting in an n^+-to-n structure. Although an additional masking step is required, these structures reduce hot electron problems that arise in short-channel length transistors. Although some of these steps may be transparent to the designer, optimizing the layout requires that one be familiar with all of the details of the processing sequence.

The drawings above have only shown the cross-sectional view of the chip. In circuit design, the important device characteristics are determined by the patterns used on each layer. Figure 2.15 shows the top view of a simple circuit after each significant processing step, illustrating the overlay properties of the patterned layers.

Since the vertical properties of the chip are specified by the fabrication line and cannot be altered, circuit design hinges on the ability to shape the layers to obtain devices with the proper characteristics.

2.5 Masks and Technologies in L-Edit

Each material layer in the IC fabrication sequence requires a separate mask that defines the geometric patterns. The top-view drawings in Fig. 2.16 show how the material patterns are overlayed during each step in the processing. Note that each layer is shown with a different fill pattern to help keep track of the patterns. L-Edit distinguishes among layers by using names and assigning a distinct color to each; fills are also used in some cases.

2.5.1 L-Edit Default Technology File

Launching L-Edit loads a pre-specified technology file into the program. This assigns a meaning and color to each layer. The default technology file is labeled **SCNA**, which stands for Scalable CMOS N-well Analog. It has all of the layers described in the sample process above, plus another polysilicon layer. Table 2.1 provides the layer names used in L-Edit's setup file.

This process illustrates a correlation between the layers and the masks used in

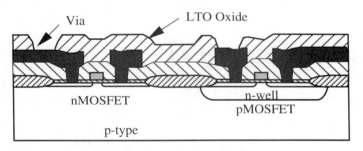

(a) LTO oxide and VIA definitions

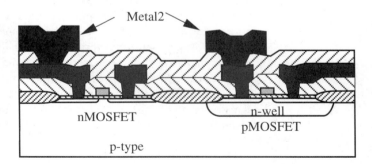

(b) Metal 2 Deposition

Figure 2.14. LTO and Metal 2 deposition.

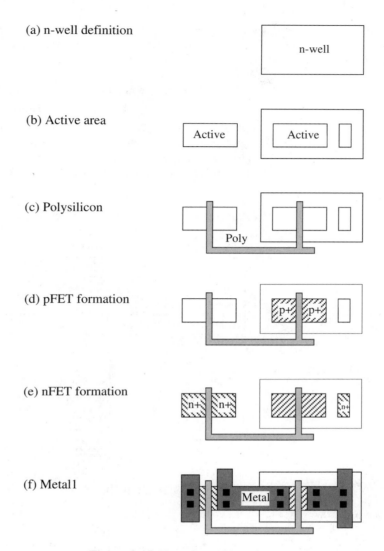

(a) n-well definition

(b) Active area

(c) Polysilicon

(d) pFET formation

(e) nFET formation

(f) Metal1

Figure 2.15. Top-view of layering sequence.

the fabrication process. However, specific device regions may or may not correspond to the pattern specified by a single mask. Consider first the METAL1 mask. It independently provides interconnect from one location to another; as such, METAL1 patterns by themselves have meaning to the final structure.

A different situation arises when describing n^+ and p^+ regions in the chip; these are obtained using a **derived layer**, i.e., a layer that is constructed by combing the basic masks. The n^+ regions in the L-Edit SCNA technology file are called **ndiff.** The ndiff layer is described by the logical operation

$$ndiff = (ACTIVE) \text{ AND } (NSELECT),$$

since an n+ region is created by performing an n-type ion implant where active areas have been defined. Physically, this implies that the ion implant beam is only effective where a thin (gate) oxide exists; the thick field oxide blocks the implant in the field regions of the wafer[1].

Similarly, p^+ regions in the n-well technology are called **pdiff** layers such that

pdiff = (ACTIVE) AND (PSELECT) AND (NWELL).

Derived layers such as ndiff and pdiff can be created using the Generate Layers option under the Special heading in the Menu Bar. The derived layers are automatically generated when L-Edit executes the Extract or Cross-Sectional viewer features, making their usage very simple.

TABLE 2. 1 Basic SCNA CMOS Layers

Physical Layer	NAME	COLOR
n-well	NWELL	TAN Outline
Silicon nitride	ACTIVE	GREEN Filled
Polysilicon Layer 1	POLY1	RED Filled
Polysilicon Layer 2	POLY2	TAN Filled
p+ Ion Implant	PSELECT	GREY Box
n+ Ion Implant	NSELECT	BLUE Outline
Contact cut to n+/p+	CONTACT TO ACTIVE	BLACK Filled
Metal 1	METAL1	BLUE Filled
Metal 2	METAL2	GREY Filled
Via oxide cuts	VIA	WHITE Filled
Pad contacts (overglass)	OVERGLASS	PURPLE Crosshatch

2.5.2 Layout Drawings

Layout drawings provide a "top view" of the chip layers, with each layer represented by a distinct color and/or fill pattern. If we draw the layers in order starting from the n-well definition, then it is possible to visualize each step in the fabrication sequence. However, a layout editor does not require that we draw the patterns in order. Instead, the sequencing information is stored internally, and any layer can be drawn at any time.

L-Edit allows you to draw any polygon on any layer at any time. For example, a basic n-channel MOSFET requires

- ACTIVE
- NSELECT
- POLY

---

[1] Many L-Edit features are based on this type of physical consideration.

CMOS Fabrication

to form the gate and the ndiff regions, but these can be drawn in any order. This feature not only makes the original drawing much easier, but also makes editing an existing cell straightforward.

Integrated circuit layout is based on the concept of creating objects with specific size and shape on each layer, and then "stacking" the layers to create 3-dimensional devices. L-Edit is designed for ease of use, and creating devices such as MOSFETs is very straightforward. Moreover, the L-Edit Cross-Sectional Viewer feature allows you to visualize the chip layers and patterns for any section of the layout.

2.6 Using Other Technologies

Launching this version of L-Edit automatically loads the database for a 2-micron, n-well CMOS process (unless the file ledit.tdb has been modified). The technology is denoted as SCNA (Scalable CMOS N-well Analog), and is contained in the disk files **ledit.tdb** and **morbn20.tdb**. This technology will be used in Chapters 3-6 for all discussions and examples that deal with CMOS integrated circuits. However, L-Edit is a basic layout editor, and is not restricted to any particular data base.

L-Edit allows you to use a predefined technology base, or create a custom data base that characterizes another fabrication process. Tanner Research has included the technology description for several CMOS processes currently accessed by universities. These are described by the technology files listed below.

MORBN20.TDB	The technology setup file for MOSIS's Orbit Semiconductor n-well 2.0 μm CMOS process. Technology = SCNA, Lambda = 1.0 μm.
MORBP20.TDB	The technology setup file for MOSIS's Orbit Semiconductor p-well 2.0 μm CMOS process. Technology = SCPE, Lambda = 1.0 μm.
MVTIN20.TDB	The technology setup file for MOSIS's VLSI Technology n-well 2.0 μm CMOS process. Technology = SCN, Lambda = 1.0 μm.
ORBTN12.TDB	The technology setup file for Orbit Semiconductor's n-well 1.2 μm CMOS process. Technology = N122P2M, Rules = MOSIS_12.
ORBTN16.TDB	The technology setup file for Orbit Semiconductor's n-well 1.6 μm CMOS process. Technology = N162P2M, Rules = MOSIS_16.
ORBTN20.TDB	The technology setup file for Orbit Semiconductor's n-well 2.0 μm CMOS process. Technology= N202P2MNPNBCCD, Rules = MOSIS_16.
ORBTP12.TDB	The technology setup file for Orbit Semiconductor's p-well 1.2 μm CMOS process. Technology = P122P2M, Rules = MOSIS_12.
ORBTP16.TDB	The technology setup file for Orbit Semiconductor's p-well 1.6 μm CMOS process. Technology = P162P2M, Rules = MOSIS_16.

ORBTP20.TDB   The technology setup file for Orbit Semiconductor's p-well 2.0 μm CMOS process. Technology = P202P2M, Rules = MOSIS_20.

MHP_N12.TDB   The technology setup file for MOSIS's Hewlett-Packard n-well 1.2 μm CMOS process. Technology = SCN, Lambda =0.6 μm.

MHP_N16.TDB   The technology setup file for MOSIS's Hewlett-Packard n-well 1.6 μm CMOS process. Technology = SCN, Lambda = 0.8 μm.

On the L-Edit disk, these files are stored in the TECH subdirectory.

To load a different technology file while L-Edit is running, choose **Replace Setup** from the **File** window on the Menu Bar. This results in the dialog box shown in Figure 2.16. Point to the box in the upper portion of the box, and enter the name of the file that you want to load. If you have not copied the file into the root directory, then you must include a path in the name. Clicking on **OK** loads the new database. L-Edit will redraw the entire screen with the new technology description in the Technology Palette. All cells and files created with the new technology base will be saved with the new information.

If you would like to configure L-Edit to automatically load a different technology file when the program is launched, just copy the technology base into the file named ledit.tdb. For example, the command

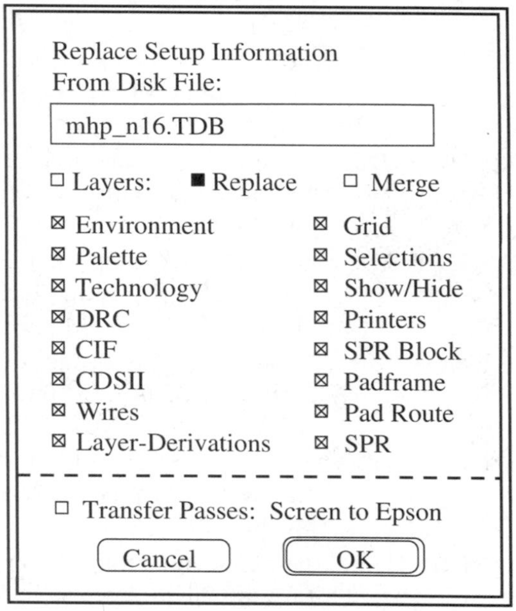

Replace Setup Information
From Disk File:

mhp_n16.TDB

☐ Layers: ■ Replace   ☐ Merge

☒ Environment   ☒ Grid
☒ Palette   ☒ Selections
☒ Technology   ☒ Show/Hide
☒ DRC   ☒ Printers
☒ CIF   ☒ SPR Block
☒ CDSII   ☒ Padframe
☒ Wires   ☒ Pad Route
☒ Layer-Derivations   ☒ SPR

– – – – – – – – – – – – – – – – – – –
☐ Transfer Passes: Screen to Epson

[Cancel]   [OK]

Figure 2.16.
The setup dialog window.

will change the default technology to the MOSIS Hewlett-Packard N-well 1.2 micron process. It is recommended that all of the technology files provided on your L-Edit disk be kept intact. In particular, **do not rename** any technology base file; use the copy operation instead.

Once you have loaded a new technology file into L-Edit, all new layout drawings will be based on the new information. Saving a file automatically saves the technology file, so that every time you load a layout drawing, you are using the same technology base that was used to create the design. As will be discussed in later chapters, the L-Edit operations of Extract and Cross-Section depend on the choice of technology base. If you change the technology, then you must choose consistent data files for these features.

The data provided on the L-Edit disk were correct when this book was written. However, process technologies are always subject to updates and new features. If you are creating a design for eventual submission to a foundry for fabrication, be sure to check to ensure that the data are consistent with the current process.

2.7 Process-Induced Failure Mechanisms

The process flow discussed above assumes ideal conditions and results. In a realistic fabrication environment, process variations and non-ideal behavior are unavoidable. Since even a single defect in an integrated circuit renders a non-functional chip, failure mechanisms are extremely important to the economic viability of the industry. Although a detailed treatment is beyond the scope of this book, it is worthwhile to examine two of the more important problem areas.

We will conclude this chapter by mentioning two failure mechanisms (of several) that are common in CMOS. These were chosen because they illustrate how the fabrication process may yield structures that do not have the ideal characteristics of the layouts.

Gate-Oxide Shorts

Gate-oxide shorts (GOS) are defects that are induced while growing the thin gate oxides needed in the MOSFET structure. The silicon dioxide layer is created by passing oxygen over the surface of the silicon via the reaction

$Si + O_2 \rightarrow SiO_2$.

A point defect at the silicon surface may cause a different growth rate than in the surrounding area, resulting in a "pinhole" in the oxide; this is illustrated in Figure 2.17(a). Since the gate polysilicon is deposited over the gate oxide, a gate-oxide short is formed through the pinhole [see Figure 2.17(b)]. Typically, the poly is doped n-type to reduce the parasitic resistance. In this case, the GOS creates a pn junction diode as modelled in Figure 2.17(c). Obviously, this will not function correctly as a MOSFET, making this type of failure mechanism a particularly important one.

If a p-channel MOSFET has a GOS, two possible situations may arise. Recall that a pFET is contained in an n-type well region that acts as the bulk terminal for the transistors. If the poly gate is doped p-type, then a GOS will induce a pn junc-

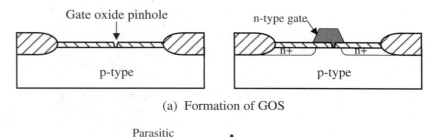

(a) Formation of GOS

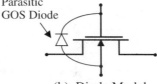

(b) Diode Model

Figure 2.17. Gate-oxide shorts (GOS).

tion. On the other hand, if the poly is doped n-type, then an n-POLY to n-well contact is formed, giving an electrical short circuit.

Problems with gate-oxide shorts tend to increase as the oxide thickness x_{ox} is decreased, since uniform, defect-free thin oxides are more difficult to grow. High-performance transistors require small values of x_{ox}, so GOS-related failures are an important consideration in state-of-the-art designs.

Line Breaks

Another important failure mechanism that arises in any type of high-density circuit is a line break. The problem is illustrated with the drawing shown in Figure 2.18. A break of this type typically arises from a defect in the mask, a speck of dust, a problem with the printing, or non-uniform material deposition. Regardless of the source, the effect is the same: an electrical open circuit.

Line break

Figure 2.18.
Defect due to a
line break.

2.8 References

There are many excellent textbooks that deal with the intricacies of integrated circuit fabrication. The ones listed below provide a sampling of the available literature.

R2.1 W. Maly, **Atlas of IC Technologies**, Benjamin-Cummings, Menlo Park, CA, 1987. This book provides excellent top-view and three-dimensional perspective drawings of nMOS and CMOS chips.

R2.2 J.Y. Chen, **CMOS Devices and Technologies for VLSI**, Prentice-Hall, Englewood Cliffs, NJ, 1990.

R2.3 W.R. Runyan and K.E. Bean, **Semiconductor Integrated Circuit Processing Technology**, Addison-Wesley, Reading, MA, 1990. A general treatment of silicon processing techniques.

R2.4 C. Mead, **Analog VLSI and Neural Systems**, Addison-Wesley, Reading, MA, 1989. Although this book is primarily directed toward analog circuits that implement neural functions, it contains an excellent summary of CMOS fabrication.

R2.5 S.K. Ghandhi, **VLSI Fabrication Principles**, 2nd ed., John Wiley & Sons, New York, 1994.

R2.6 M. Shoji, **Theory of CMOS Digital Circuits and Circuit Failures**, Princeton University Press, Princeton, NJ, 1992.

R2.7 J.P. Uyemura, **Circuit Design for CMOS VLSI**, Kluwer Academic Publishers, Norwell, MA, 1992.

R2.8 J.P. Uyemura, **Fundamentals of MOS Digital Integrated Circuits**, Addison-Wesley, Reading, MA, 1988.

Chapter
3

Layout of CMOS Integrated Circuits

To design a digital integrated circuit, one usually starts with a circuit schematic. This provides the topology of the network that implements the logic. Layout design is the next step in the process. It is directed toward the problem of translating the schematic into a set of patterned layers that form the integrated structure in a silicon substrate. All of the electrical properties of the circuit are established in this phase of the design sequence.

This chapter examines the concepts and details used to implement CMOS circuits in silicon.

3.1 Introduction to Physical Design

Several equivalent viewpoints may be used to describe an integrated circuit. To a circuit designer, a chip is the physical realization of an electronic network. A logic designer, on the other hand, may choose to view the chip as a device that performs functions specified by logic diagrams, function tables, or an HDL[1] file. Figure 3.1 illustrates how different people might view the same thing. Regardless of the abstraction used, in the final analysis, an integrated circuit is really an intricate physical object that has been carefully designed and fabricated.

Physical design in VLSI deals with the procedure needed to realize a circuit on the surface of a semiconductor wafer. Starting with the electrical network schematic, computer tools are used to create the necessary patterns on each layer in the 3-dimensional structure. Once the drawings are completed, the information can be used to fabricate the masks needed in the processing line.

---

[1] HDL is an acronym that stands for **hardware description language**.

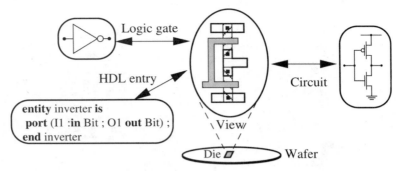

Figure 3.1. Equivalent descriptions of a digital integrated circuit.

CMOS technology allows one to choose from a wide variety of circuit design techniques, any of which may be useful when implementing a given logic function. This feature is particularly nice when designing high-performance circuits, as often one design style yields faster switching than another. Physical design is critical in this situation, since the layout and the resulting performance are directly linked to each other. At this level, the circuit design and layout are indistinguishable.

Many people view physical design as a skill that is best learned by doing. The most proficient designers tend to be the most experienced, but, of course, one must begin somewhere. In this chapter, we will introduce the first ideas of physical design by examining the concept of layout in more detail. This includes ideas such as design rules and interconnect routing. The details of designing CMOS circuits will be covered in the following chapters.

3.2 Masks and Layout Drawings

Every material layer in an integrated circuit is described by a set of geometrical objects of specified shape and size. These objects are defined with respect to each other on the same layer, and also with reference to geometrical objects that lie on other layers, both above and below. Layout drawings relay this information graphically, and can be used to generate the masks needed in the fabrication process. Because of this relationship, we will take the viewpoint that a layout drawing represents the top view of the chip itself.

When we visualize an integrated circuit, it appears as a set of overlapping geometrical objects. In a layout editor, each layer is described by using a distinct color or fill pattern that allows us to see the objects relative to each other. Once we get oriented to seeing an integrated circuit in this manner, it is a simple matter to construct transistors and route the interconnect lines as required. While classical schematic representations provide the topology of the network, the layout gives us the ability to modify the performance of a circuit. Performing the layout is therefore an intrinsic part of the design process.

When designing digital logic circuits in CMOS, the goals remain quite simple:

- Design a circuit that implements the logic function correctly, and,

- Adjust the parameters to meet the electrical specifications.

This is often more difficult than it sounds, particularly when we note that state-of-the art VLSI chips can have several million MOSFETs with the associated interconnect lines. At the most basic level, we find that many problems arise when performing the layout of an integrated circuit. Some deal with the practical aspects of circuit operation, others originate from physical properties of the materials involved, and yet others are due to limitations in the fabrication processes. These all contribute to the techniques used in the physical design.

3.3 Design Rule Basics

Design rules are a set of guidelines that specify the minimum dimensions and spacings allowed in a layout drawing. They are derived from constraints imposed by the processing and other physical considerations. Violating a design rule may result in a non-functional circuit[1], so that they are crucially important to enhancing the die yield.

Limitations in the photolithography and pattern definition give rise to several critical design rules. Since these are strongly dependent on equipment used in the fabrication process, they tend to change with improving technology. The situation is complicated by the fact that physical phenomena and device design considerations also enter into the picture. In this section, we will examine some of the design rules associated with a CMOS processing technology.

3.3.1 Minimum Linewidths and Spacings

Consider the two objects shown in Figure 3.2. These represent two patterns on the same layer, e.g., both are polysilicon. When used as interconnects, the two rectangles shown in the drawing are called "lines" (since every physical patterning must have a non-zero width), and we will use this terminology in our discussion. The **minimum linewidth** X is the smallest dimension permitted for any object in the layout drawing; X is also known as the **minimum feature size**. The **minimum spacing** S is the smallest distance permitted between the edges of two objects; in the present example, the minimum spacing is between the two lines.

Minimum linewidth and spacing values for interconnect lines may originate from the resolution of the optical printing system, the etching process, or from other considerations such as surface roughness. Violating the minimum linewidth rule may result in ill-defined or broken interconnects. Similarly, the minimum spacing rule ensures that the lines are physically separated in the final structure. If this guideline is not followed, then the two may form an electrical short in the circuit.

The situation is more complicated when applied to patterning a doped region in the silicon because of lateral doping effects and the presence of depletion regions. Lateral doping is shown in Figure 3.3. In (a), the oxide has been patterned by the lithography to have a linewidth X. However, when the wafer is heated to anneal the implant, lateral diffusion[2] increases the actual width of the n^+ region to $X'>X$. This

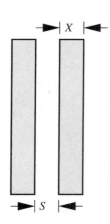

Figure 3.2.
Minimum width and spacing.

---

[1] Note that even a single defect results in a non-functioning circuit.

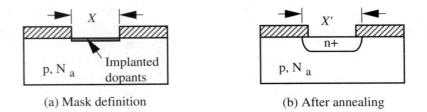

(a) Mask definition   (b) After annealing

Figure 3.3. Patterning sequence for a doped n^+ line.

effect is important when determining the minimum spacing S between adjacent doped lines.

Depletion effects also influence the value of S. As shown in Figure 3.4, a depletion region exists at every pn junction. Let us assume for simplicity that the junction has a step-doping profile where the impurity concentration changes abruptly from N_d on the n-side to N_a on the p-side. With a reverse-bias voltage of V_R, the total depletion width x_d can be computed from

$$x_d = x_0 \sqrt{1 + \frac{V_R}{V_{bi}}},$$

(3. 1)

where

$$x_0 = \sqrt{\frac{2\varepsilon_{Si} V_{bi}}{q}\left(\frac{1}{N_a} + \frac{1}{N_d}\right)}$$

(3. 2)

is the zero-bias value of the total depletion width, and

$$V_{bi} = \left(\frac{kT}{q}\right)\ln\left(\frac{N_a N_d}{n_i^2}\right)$$

(3. 3)

is the built-in voltage. Table 3.1 provides a list of useful numerical values for basic calculations. Note that the intrinsic concentration n_i applies only to silicon at room temperature ($T=300°\ K$). To calculate the p-side depletion width x_p shown in the drawing, we use

$$x_p = \left(\frac{N_d}{N_d + N_a}\right)x_0 \sqrt{1 + \frac{V_R}{V_{bi}}}.$$

(3. 4)

Since this increases with the reverse bias voltage, the minimum spacing distance S often accounts for the worst-case situation, i.e., when $V_R=V_{DD}$. From this discussion, it is not surprising that the minimum width and spacing for n^+ and p^+ regions are usually larger than those for a polysilicon line.

---

[2] "Lateral" means in a direction parallel to the surface.

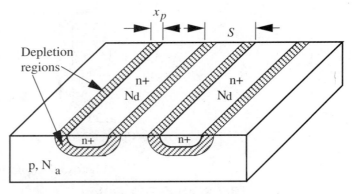

Figure 3.4. Depletion regions due to parallel n$^+$ lines.

3.3.2 Contacts and Vias

Contacts and vias are used to provide electrical connections between different material layers. In general, contacts are necessary connections to access the various regions of silicon, while vias are used between two interconnect layers to simplify the layout. When formulating design rules for these types of objects, two important considerations arise: the physical size of the oxide cuts, and the spacing needed around the connection on the layers.

TABLE 3.1 Useful constants

Symbol	Parameter name	Value	Units
q	Electron charge	1.6×10^{-19}	Coulombs
$\varepsilon_{Si}=\varepsilon_r\varepsilon_0$	Permittivity of silicon	$(11.8)(8.854 \times 10^{-14})$	Farads/cm
$\varepsilon_{ox}=\varepsilon_r\varepsilon_0$	Permittivity of SiO$_2$	$(3.9)(8.854 \times 10^{-14})$	Farads/cm
n_i	Intrinsic concentration	1.45×10^{-10}	cm^{-3}
(kT/q)	Thermal voltage	0.0258	Volts

Let us first examine the dimensions of a contact region. The geometry is shown in Figure 3.5. It is apparent that the minimum size is limited by the lithographic process. However, this does not imply that one uses the largest contacts possible, as other considerations come into play. If the contact cut is too large, then it may be difficult to attain complete coverage when depositing the upper layer. Large contact cuts may result in cracks or voids, that may in turn lead to a circuit failure. To overcome this problem, it is common to restrict the dimensions of contact cuts to pre-specified values that can be reliably made in the fabrication process.

Now, consider the problem of spacing x around an oxide cut as shown in Figure 3.6. We must specify the minimum distance between the edge of an oxide cut and the edge of a patterned region to allow for misalignment tolerances in the masking steps. These are generically classified as **registration errors**, and arise because it is not possible to align the mask with arbitrary precision.

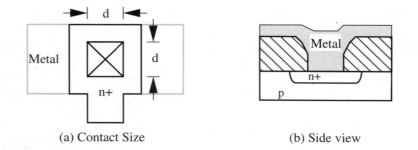

(a) Contact Size   (b) Side view

Figure 3.5. Geometry of a contact cut.

3.3.3 MOSFET Rules

MOSFETs are usually fabricated using the self-aligned technique described in the last chapter. This approach uses the polysilicon gate as a mask for the ion implantation step that forms the drain/source regions. Certain precautions must be taken in the physical design to ensure that small registration errors can be tolerated, and functional transistors will still be formed.

First, we recall that n^+ regions in an nFET are described by the derived layer[1]

$$ndiff = (ACTIVE) \text{ AND } (NSELECT).$$

With regard to the fabrication sequence, this means that an n^+ region is formed where (a) the NSELECT mask gives an opening in the photoresist, AND, (b) an ACTIVE area exists to allow the implant to penetrate into the silicon substrate. Since the formation of the physical ndiff layer relies on the overlap of two masks, the size of the NSELECT region must be larger than the size of the corresponding ACTIVE area.

The drawing in Figure 3.7(a) shows the proper sizing of the two layers, with the NSELECT rectangle larger than the ACTIVE area rectangle. The final dimensions of the n^+ region are those of the ACTIVE area. A minimum spacing value x between the edges of the two masks is used to allow for registration error between

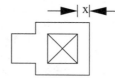

(a) Mask Design   (b) Registration
Tolerance

Figure 3.6. Contact spacing rule.

---

[1] See Section 2.5 of the previous chapter.

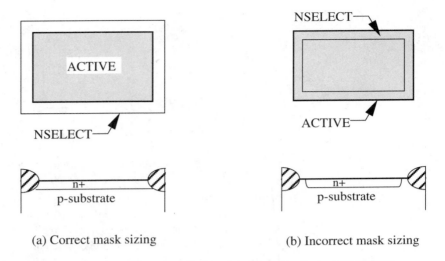

(a) Correct mask sizing   (b) Incorrect mask sizing

Figure 3.7. Formation of n^+ regions in an n-channel MOSFET

the two masks. Even though the ion implant specified by the NSELECT boundary is larger than needed, the thick field oxide specified by the region NOT(ACTIVE) prevents the underlying silicon from being doped. This is verified by the cross-sectional view shown.

If we reverse the situation and make the ACTIVE area larger than the NSELECT area, then only the region described by ndiff=(ACTIVE) AND (NSELECT) will become n^+; sections corresponding to

$$(\text{ACTIVE}) \text{ AND } [\text{NOT(NSELECT)}]$$

remain p-type substrate. This is shown in Figure 3.7(b). The ndiff region does not have the correct cross-sectional pattern needed to ensure proper isolation and operation.

Consider next the **gate overhang** distance d shown in Figure 3.8. This is included to ensure that a misaligned gate will still yield a structure that has separate drain and source regions. To understand the reasoning, suppose that a MOSFET is designed using the layout shown in Figure 3.9(a) with d=0. Figure 3.9(b) shows a

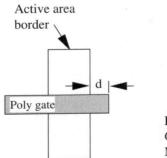

Figure 3.8.
Gate overhang in
MOSFET layout.

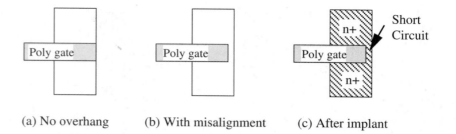

| (a) No overhang | (b) With misalignment | (c) After implant |

Figure 3.9. Effect of gate misalignment without overhang.

small misalignment where the polysilicon does not traverse the entire active area. Since the ion implant will dope all of the exposed substrate, the resulting structure shown in (c) has the drain and source n^+ regions merged into one. Electrically, the drain and source are shorted, so the device cannot control the current flow, i.e., the switching action has been lost.

The same consideration applies to a MOSFET where the n+ region changes shape as shown in Figure 3.10. The channel width W is a critical design parameter, so that the spacing *s* between the poly and n^+-edges must be large enough to ensure that the MOSFET still has the proper value if small registration errors occur.

3.3.4 Bloats and Shrinks

The drawings produced by a layout editor provide the basic view of an integrated circuit that are used to extract the equivalent device parameters. It is therefore important to understand the correlation between what is shown on the computer monitor when compared with the actual die after fabrication.

In general, the final size of a physical layer will be different from the dimensions specified by the mask that created the pattern. Two obvious examples are

- ACTIVE: Encroachment reduces the size of the usable active area.
- Doped n^+ or p^+: Lateral doping effects increase the size of these regions.

In addition, the physical process of etching a material layer is anisotropic, with both vertical and lateral etching present. Although the lateral etch rate can be reduced

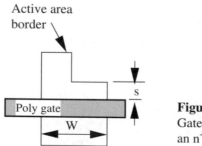

Figure 3.10.
Gate spacing from an n^+ edge.

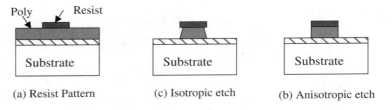

(a) Resist Pattern   (c) Isotropic etch   (b) Anisotropic etch

Figure 3.11. Polysilicon etch profiles.

using various techniques, it cannot in general be reduced to zero. The effect of anisotropic etching on a polysilicon layer is illustrated in Figure 3.11.

The question that naturally arises at this point is "What does the layout drawing represent relative to the finished chip?" In other words, will the chip patterns be identical to those shown by the layout editor, or are size adjustments necessary? In the early days of chip design, one had to increase or decrease the size of the layout drawing to view the actual chip dimensions. However, it is now common for the chip fabricator to subject the masks to bloats (increases in object sizes) and shrinks (decreases in the object size) as needed to compensate for the difference between the mask dimensions and the resulting size on the chip. When this is done, then the layout editor displays a reasonably accurate view of the finished circuit.

These considerations are particularly important to designing a MOSFET. Although the two critical dimensions L (the channel length) and W (the channel width) are related to the **drawn** mask values, the values are different as shown in Figure 3.12. The channel length L that is required in the current flow equations is reduced from the drawn value L' by

$$L = L' - 2L_o \qquad (3.5)$$

where L_o is the overlap distance from lateral diffusion effects[1]. In a similar manner, the channel width W is smaller than the drawn ACTIVE width W' because of active area **encroachment**. This is where the usable size of the active area is reduced because of oxide growth underneath the edges of the nitride pattern. If the encroachment per side is (ΔW), then

$$W = W' - 2(\Delta W) \qquad (3.6)$$

gives the proper electrical value. This can become confusing when entering the device dimension into a circuit simulation program. For example, SPICE parameters can be entered using either the drawn or physical values so long as the remaining data values are consistent. This will be discussed in more detail in the next chapter.

---

[1] The gate overlap L_o is also known as the lateral diffusion length L_D.

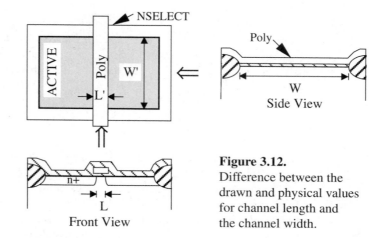

Figure 3.12.
Difference between the drawn and physical values for channel length and the channel width.

3.4 Types of Design Rules

Geometrical design rules are a set of minimum widths, spacings, and layout guidelines needed to create the masks. There are two ways to specify these dimensions:

- **Specific Values**: All dimensions are stated in standard unit of length, such as the micron;

- **Scalable**: Distances are specified as multiples of a metric λ that has dimensions of length. The actual value of λ is adjusted to correspond to the limitations of the process line.

Both approaches are common in CMOS VLSI. Scalable rules have the advantage that they can be adjusted to several different processing lines by changing the value of λ. However, this does not come without cost. Since every distance is specified as a multiple of λ, the numerical value is dictated by the worst-case situation. This generally decreases the compaction density of the circuit compared to what is attainable if the parameters are specified in an absolute metric such as microns.

In general, there are three main classes of design rule specifications. These are

- Minimum Width,
- Minimum Spacing, and,
- Surround.

Surround rules apply to objects placed within larger objects (such as contacts). Every layer has a minimum width and minimum spacing value, while surrounds are specified as required.

3.5 CMOS Design Rules

In this section, we will examine a basic set of CMOS design rules to understand the presentation and meaning of each type of rule. This set has been provided in the setup technology file as **ledit.tdb,** and is also available with the name **morbn20.tdb**. The DR set describes the MOSIS Orbit 2-micron double-poly, double-metal, n-well CMOS process; the technology name for this process is **SCNA** (for Scalable CMOS N-Well Analog). For the purposes of this discussion, we will not list all layers. In particular, the POLY2 layer is not shown explicitly here to simplify the discussion.

A complete design rule set contains all of the geometric limits for mask layout. This includes the minimum feature sizes and minimum spacings on each mask, and also provides layer-to-layer spacings when necessary. In order to list the rules in an easy-to-find format, they are presented according to the order of the masks used in the processing. The primary design rule layers for the morbn20 technology are listed in Table 3.2.

TABLE 3.2 SCNA Design Rule Layers

Mask Number	Mask Layer
1.	NWELL
2.	ACTIVE
3.	POLY
4.	SELECT
5.	POLY CONTACT
6.	ACTIVE CONTACT
7.	METAL1
8.	VIA
9.	METAL2
10.	PAD
11.	POLY2

When you are using L-Edit, the design rules corresponding to the technology are always loaded into your file, and are Saved when you save your work. A text listing of the design rules can be obtained using the keyboard command **Alt-W**; there is no Menu Bar equivalent. This action writes a text file named **filename.rul** to the working disk drive that provides a listing of all layers and rules in ASCII format. The list also provides information on derived layers.

Design rule sets are most easily understood by providing a text list in conjunction with simple drawings to illustrate each value. These are broken into groups corresponding to each basic layer. The layer number N. is used to identify the group, and each dimensional specification is labelled accordingly, e.g., N.1, N.2, and so on. In order to clarify some of the fine points involved with design rules, they will be presented in two different forms. First, we will examine a simplified set of rules that provide basic information on minimum widths and spacings, and then look at MOSFET layout rules. This gives a general idea of what the rules mean.

This is followed by a more complete set of layout statements that correspond to those used by L-Edit in performing the DRC algorithm.

3.5.1 Basic Rules

The most fundamental layout guidelines limit the smallness of each material layer, and provide the basis for device design. The values in this design rule set are scalable according to the metric λ. Numerically, $\lambda = 1$ μm for the 2 μm technology. However, these rules also apply to the mhp_n12.tdb (named **SCN**) 1.2-micron, single-poly, double-metal process with a metric of value $\lambda = 0.6$ μm.

Minimum Widths and Spacings

This group of rules are those that specify the minimum linewidths and minimum spacings permitted on the primary layers summarized in Table 3.3 and illustrated in Figure 3.13.

TABLE 3.3 Minimum Width and Spacing Rules

Layer	Type of Rule	Value
POLY	Minimum width	2λ
	Minimum spacing	2λ
ACTIVE	Minimum width	3λ
	Minimum spacing	3λ
NSELECT	Minimum width	3λ
	Minimum spacing	3λ
METAL1	Minimum width	3λ
	Minimum spacing	3λ
METAL2	Minimum width	3λ
	Minimum spacing	4λ

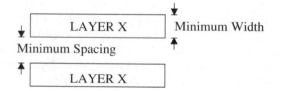

Figure 3.13.
Minimum width and minimum spacing.

MOSFET Layout

The basic layout rules for MOSFETs are illustrated in Figure 3.14, and the values are summarized in Table 3.4. Dimensions that deal with POLY and N+/P+ apply equally to both nFETs and pFETs. Contacts (cuts in the oxide that allow electrical connections between two layers) to N+/P+ are called ACTIVE CONTACTS, while POLY CONTACTS provide access to the POLY layer (which must be in field regions).

In this technology, p-channel MOSFETs must be inside n-well regions, and sufficient spacing must be provided between opposite-polarity regions (i.e., between n^+ and p^+ sections) as well as between P+ regions and the NWELL edge. Spacing requirements also apply between different n-well regions; in general, a larger space is needed if the n-wells are biased at different voltages, due to the voltage dependence of depletion regions.

TABLE 3.4 MOSFET Layout Rules

RULE	Meaning	Value
POLY Overlap	Minimum extension over ACTIVE	2λ
POLY-ACTIVE	Minimum Spacing	1λ
MOSFET Width	Minimum N+/P+ MOSFET W	3λ
ACTIVE CONTACT	Exact Size	$2\lambda \times 2\lambda$
ACTIVE CONTACT	Minimum Space to ACTIVE Edge	2λ
POLY CONTACT	Exact Size	$2\lambda \times 2\lambda$
POLY CONTACT	Minimum Space to POLY Edge	2λ

3.5.2 Mask-Based Design Rule Set

At the layout level, design rules apply to the masks involved in the chip patterning process. L-Edit employs a complete set of rules that apply to both masking layers and derived layers. Using derived layers allows us to make the connection between the individual patterns and the material layers that are important to the operation of transistors and other devices.

Figures 3.15, 3.16, and 3.17 provide most of the SCNA design rule set using

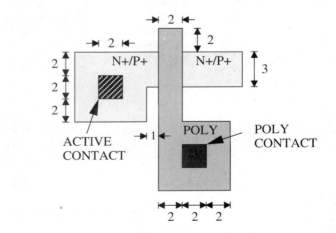

Figure 3.14.
MOSFET
layout rules.

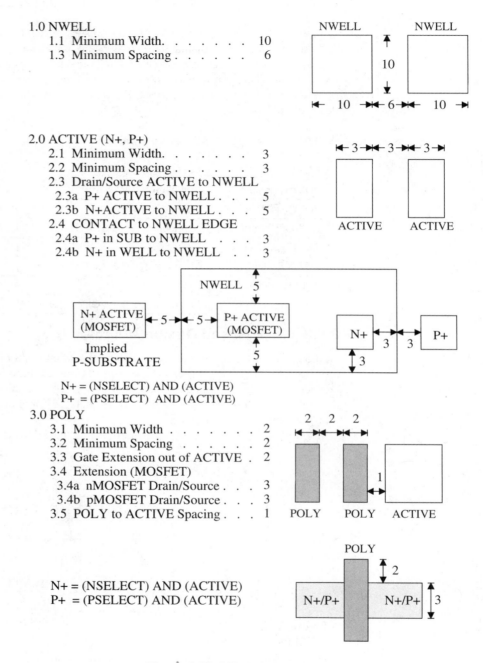

1.0 NWELL
    1.1 Minimum Width. 10
    1.3 Minimum Spacing 6

2.0 ACTIVE (N+, P+)
    2.1 Minimum Width. 3
    2.2 Minimum Spacing 3
    2.3 Drain/Source ACTIVE to NWELL
    2.3a P+ ACTIVE to NWELL . . . 5
    2.3b N+ACTIVE to NWELL . . . 5
    2.4 CONTACT to NWELL EDGE
    2.4a P+ in SUB to NWELL . . . 3
    2.4b N+ in WELL to NWELL . . 3

N+ = (NSELECT) AND (ACTIVE)
P+ = (PSELECT) AND (ACTIVE)

3.0 POLY
    3.1 Minimum Width 2
    3.2 Minimum Spacing 2
    3.3 Gate Extension out of ACTIVE . 2
    3.4 Extension (MOSFET)
    3.4a nMOSFET Drain/Source . . . 3
    3.4b pMOSFET Drain/Source . . . 3
    3.5 POLY to ACTIVE Spacing . . . 1

N+ = (NSELECT) AND (ACTIVE)
P+ = (PSELECT) AND (ACTIVE)

Figure 3.15. SCNA design rules.

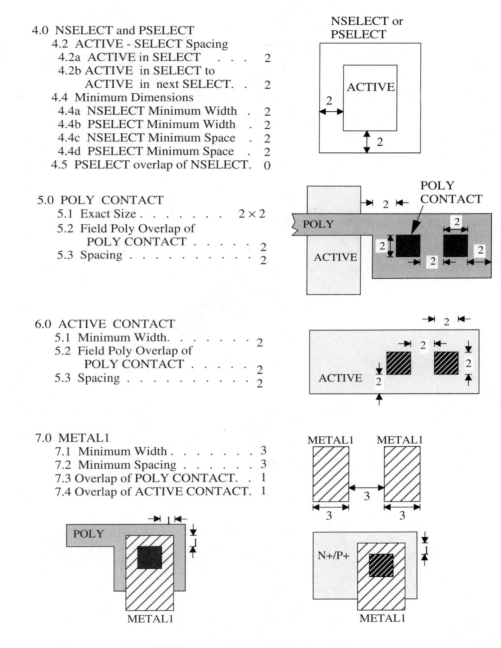

4.0 NSELECT and PSELECT
    4.2 ACTIVE - SELECT Spacing
    4.2a ACTIVE in SELECT . . . 2
    4.2b ACTIVE in SELECT to
          ACTIVE in next SELECT. . 2
    4.4 Minimum Dimensions
    4.4a NSELECT Minimum Width . 2
    4.4b PSELECT Minimum Width . 2
    4.4c NSELECT Minimum Space . 2
    4.4d PSELECT Minimum Space . 2
    4.5 PSELECT overlap of NSELECT. 0

5.0 POLY CONTACT
    5.1 Exact Size . . . . 2 × 2
    5.2 Field Poly Overlap of
          POLY CONTACT 2
    5.3 Spacing 2

6.0 ACTIVE CONTACT
    5.1 Minimum Width. 2
    5.2 Field Poly Overlap of
          POLY CONTACT 2
    5.3 Spacing 2

7.0 METAL1
    7.1 Minimum Width 3
    7.2 Minimum Spacing 3
    7.3 Overlap of POLY CONTACT. . 1
    7.4 Overlap of ACTIVE CONTACT. 1

Figure 3.16. SCNA design rules (continued).

8.0 VIA
  8.1 Exact Size . . . .   2 × 2
  8.2 VIA to VIA Spacing. 3
  8.3 METAL1 Overlap of VIA . . 1
  8.4 VIA Spacing
   8.4a VIA to POLY. 2
   8.4b VIA (on POLY) to POLY . . 2
   8.4c VIA to ACTIVE 2
   8.4b VIA (on ACTIVE) to POLY . 2

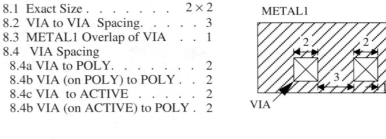

9.0 METAL2
  9.1 Minimum Width 3
  9.2 Minimum Spacing 4
  9.3 Overlap of VIA 4

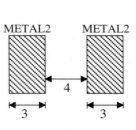

Figure 3.17. SCNA design rules (continued).

values in conjunction with drawings. Some have been omitted or merged for simplicity, but can be accessed by having L-Edit print the complete set. In the drawings, note that the derived layers

$$N+ = (NSELECT) \text{ AND } (ACTIVE)$$

$$P+ = (PSELECT) \text{ AND } (ACTIVE)$$

are equivalent to **ndiff** (n^+) and **pdiff** (p^+) doped regions.

3.6 Design Rules in L-Edit

Design rule sets are automatically loaded with the technology base when L-Edit is launched. Also, design rule information is saved with a work file. L-Edit allows the user to modify any or all of the rules in a set, or load an alternate set. The DRC (design rule checker) algorithm is used to analyze the layout to see if all of the rules are satisfied, and to report any violations that are found.

3.6.1 Entering Design Rule Values

You may examine the existing set, edit the parameters, or add new rules using

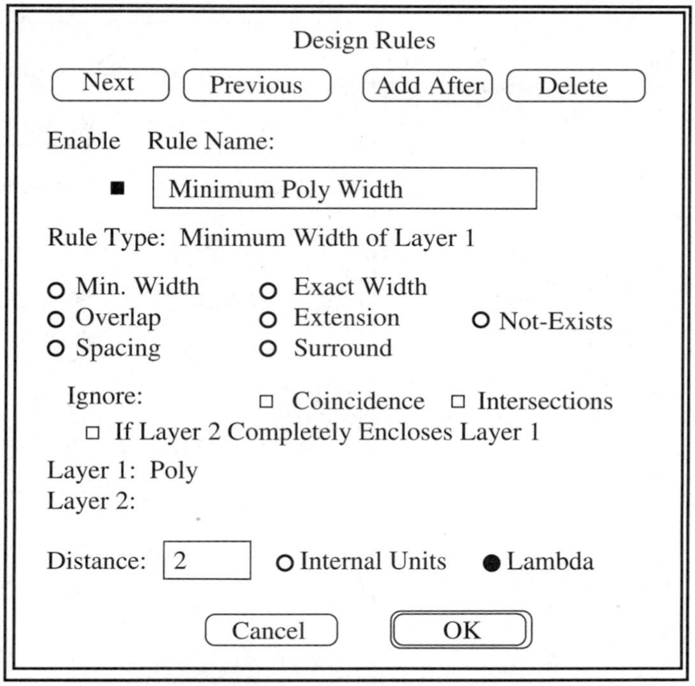

Figure 3.18. The L-Edit design rule dialog window.

the DRC command in the Layers window of the menu bar. Executing this option opens the dialog box shown in Figure 3.18.

It is important to note that L-Edit uses a pre-specified metric to interpret the distance between grid points. In a scalable technology, the units will be designated as Lambda. To check the units that are assumed, invoke the Technology command in the Setup window. This action opens the dialog box shown in Figure 3.19, which defines the distance between grid points and the value of Lambda.

3.6.2 Running a Design Rule Check

To execute a DRC on the layout, choose DRC from the Special window in the Menu Bar. This action opens a dialog box that allows you to specify the format of the output. There are three modes that can be used to obtain information on the violations: placing error ports at each location, placing error markers at each location, and/or generation of a text file listing.

The L-Edit DRC will show the location of design rule violations directly on the drawing. These are created on the Error Layer reserved for this purpose. Violations are marked by ports that are drawn on the Error Layer. To clear this layer, use the Clear Error Layer command. The Find Object command can be used to view each error by specifying the ports as the objects, and the Error Layer as the layer to be searched. You may browse the entire set using Find Next Object and Find Previous Object commands.

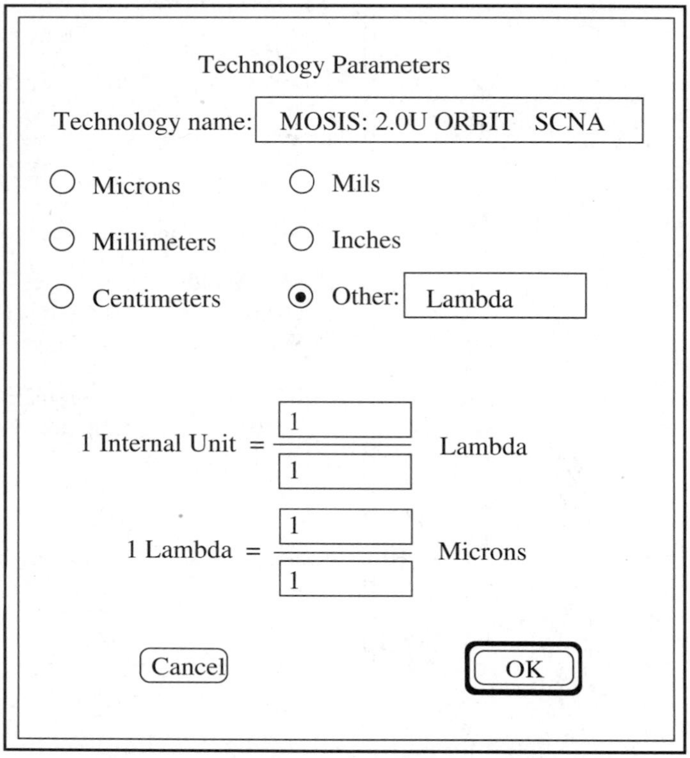

Figure 3.19. Grid spacing definition in L-Edit.

L-Edit will provide a text file listing of every violation. The format of each line in the report is as follows:

<Rule Name> Type: <Rules Type>, Distance: <distance><unit name>
Layer: <Layer Name>
Layer: <Layer Name>
Layer: <Layer Name>

where
    <Rule Name> is the name of the design rule;
    <Rule Type> is the type of rule, e.g., minimum width;
    <distance> gives the coordinates where the violation occurred in the form;
    (x1,y1) -> (x2,y2) with the "->" representing an arrow;
    <Layer Name> gives the name of the layer involved.

This allows you to quickly find the errors in the layout.

L-Edit Example

To illustrate the operation of the DRC in L-Edit, let us examine the objects shown in the screen dump reproduced in Figure 3.20. The drawing was created with several design rule violations including minimum width (POLY and METAL1), minimum spacing (POLY and METAL1), and surround (ACTIVE in NSELECT). The grid points in the plot are separated by 1λ, and the origin (0,0) has been included for reference. Running a DRC on the file results in the following text file listing:

DRC Errors in cell Cell0 of file DRC Example.
7.2 Metal1 to Metal1 Spacing = 3 Lambda: (19,0)->(17,0)
7.2 Metal1 to Metal1 Spacing = 3 Lambda: (19,20)->(17,20)
7.1 Metal1 Width = 3 Lambda: (17,0)->(15,0)
7.1 Metal1 Width = 3 Lambda: (15,20)->(17,20)
6A.2 FieldAct Overlap ActCnt = 2 Lambda: (6,4)->(12,10)
6A.2 FieldAct Overlap ActCnt = 2 Lambda: (6,8)->(12,14)
6A.2 FieldAct Overlap ActCnt = 2 Lambda: (6,12)->(12,18)
4.2.b1 Active (in Select) to Select Edge = 2 Lambda: (7,5)->(6,5)
4.2.b1 Active (in Select) to Select Edge = 2 Lambda: (24,5)->(25,5)
4.2.b1 Active (in Select) to Select Edge = 2 Lambda: (24,17)->(25,17)
4.2.b1 Active (in Select) to Select Edge = 2 Lambda: (7,17)->(6,17)
3.2 Poly to Poly Spacing = 2 Lambda: (1,1)->(2,1)
3.2 Poly to Poly Spacing = 2 Lambda: (1,17)->(2,17)
3.1 Poly Width = 2 Lambda: (1,1)->(0,1)
3.1 Poly Width = 2 Lambda: (0,17)->(1,17)
15 errors.

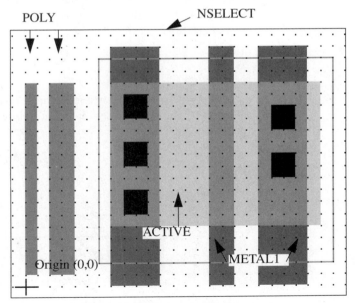

Figure 3.20. L-Edit DRC example.

As discussed above, the format of the text file indicates the rule that was violated and the coordinates where the violation was found. For example, a violation of the minimum Metal1 width rule of 3λ occurred at coordinates (17,0)->(15,0) and (15,20)->(17,20), which is the center Metal1 line shown. Each error can be located in the same way. This type of listing allows you to quickly identify the problem.

3.7 Parasitics

Parasitic elements are electrical components that arise from the structuring of a circuit due to inherent electromagnetic effects. In an integrated circuit environment, parasitic resistance and capacitance are the elements that limit the switching performance of a digital logic network. As such, it is very important to understand the origin of these elements, and how to compute them.

In general, parasitics can be split into two types, those originating from the transistors and those associated with interconnects. MOSFET contributions are discussed in detail in the next chapter, so we will confine ourselves to analyzing interconnect structures in the present discussion.

3.7.1 Resistance

Parasitic resistance in an interconnect line is due to the finite conductivity σ of the material. Consider the top-view interconnect geometry shown in Figure 3.21. The end-to-end resistance is computed from

$$R = \frac{d}{\sigma A},$$

(3. 7)

where A is the cross-sectional area of the conductor in a direction perpendicular to the current flow in the form $A = wt$, with t the thickness of the material. To simplify the calculations, we introduce the sheet resistance R_s defined by

$$R_s = \frac{1}{\sigma t} = \frac{\rho}{t},$$

(3. 8)

where $\rho = (1/\sigma)$ is the resistivity in units of $\Omega\text{-}cm$. Although R_s has strict units of ohms (Ω), we usually assign it units of **ohms per square** since the resistance of a line having a width w and a length d is given by

$$R_{line} = R_s n$$

(3. 9)

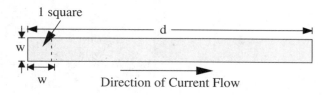

Figure 3.21. Geometry for resistance calculation.

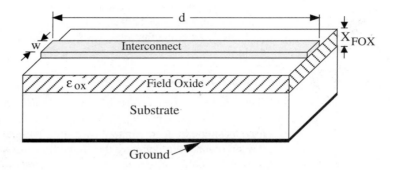

Figure 3.22. Geometry for interconnect capacitance calculation.

where $n=(d/w)$ is the "number of squares" with dimensions $(w \times w)$ as shown in the illustration. For an interconnect with a specified length d, increasing the width w decreases the line resistance. Sheet resistance values are provided in the process specifications.

3.7.2 Capacitance

Inteconnect capacitance can be classified as self-capacitance or coupling capacitance. Self capacitance is referenced to ground, while coupling capacitances are referenced to each other. Consider the geometry shown in Figure 3.22 where a metal line is patterned over a field oxide with a thickness X_{FOX}. Using the simple parallel-plate capacitor formula gives the interconnect capacitance

$$C_{line} = C_I A,$$ (3.10)

where

$$C_I = \frac{\varepsilon_{ox}}{X_{FOX}}$$ (3.11)

is the interconnect capacitance per unit area, and $A = wd$ is the area of the line. Since this equation ignores the existence of fringing electric fields, this approach tends to underestimate the actual value. However, it does give reasonable estimates for the initial calculations, and more accurate equations can be found in the literature. Note that increasing the line width w increases the line capacitance.

3.7.3 Interconnect Modelling

CMOS circuits are often performance-limited by interconnect parasitics. Typical problems include excessive line delay and timing skews. Since both the resistance and capacitance per unit length of a patterned line are proportional to the length, long lines often merit a closer look to ensure that they do not adversely affect the speed of a critical data flow path.

Various models can be used to include the effects of parasitics on the circuit design. As expected, the most accurate modes are the most complicated, so that the designer must choose between precision and simulation time. In this section, we will briefly examine three levels of interconnect models that can be used in CMOS design.

Lumped RC Model

The simplest model consists of the simple RC network shown in Figure 3.23. Here, the total line resistance R_{line} [Ω] and the total interconnect capacitance C_{line} [F] are used as single lumped-element parasitics. Although this oversimplifies the effect of the interconnect in delaying the signal transmission, it is useful for a first estimate of the line delay. In general, this model tends to give pessimistic results, i.e., the predicted delay times are longer than the actual values.

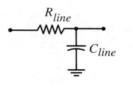

Figure 3.23.
Lumped element model
for interconnect parasitics.

RC Ladder Network

A more accurate approach is to construct an RC ladder network made up of m-rungs, each consisting of elements with values $R=(R_{line}/m)$ and $C=(C_{line}/m)$. Figure 3.24 illustrates this construction for m=4. Creating the ladder more accurately models the effects of the distributed parasitics. It can be shown that this provides a higher degree of accuracy than that obtained using a single RC segment; the accuracy improves with increasing value of m. However, since the simulation time increases with the number of nodes, this model requires longer computer times.

Distributed-Elements

The most accurate model for signal transmission is to use a distributed model where the parasitics are described by values per unit length in R' Ω/cm and C' $F/$

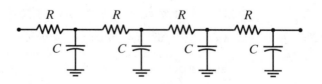

Figure 3.24. RC ladder approximation for interconnect.

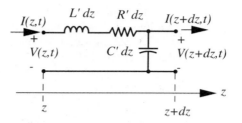

Figure 3.25. Differential transmission line model.

cm. The voltage and current then become functions of both position z and time t on the line, and are denoted by $V(z,t)$ and $I(z,t)$. Figure 3.25 shows a differential section of length dz. The inductance per unit length L' H/cm may also be included.

The transmission line analysis results in a pair of coupled partial-differential equations for the voltage and current. If L' is small, then the voltage and current "diffuse" from one end to the other at a rate that depends on the $R'C'$ product, with small $R'C'$ values giving fast signal transmission, as expected on physical grounds. The other extreme case is that where L' is large enough to give wave effects where a wavefront travels at a phase velocity of

$$v_p = \sqrt{\frac{L'}{C'}} = \frac{c}{\sqrt{\varepsilon_r}},$$

(3. 12)

where c is the speed of light in a vacuum and ε_r is the relative permittivity of the oxide[1]. For silicon dioxide, $\varepsilon_r \approx 3.9$, so $v_p \approx 152$ $[\mu m/ps]$. Since CMOS switching times are at best on the order of 100 ps, on-chip transmission line effects are not generally observed, and the lumped-element models are sufficient. However, wave properties can be important for off-chip driver networks.

General Comments

The parasitics associated with inteconnect lines are a significant limiting factor in the design of high-speed circuits. Large values of capacitance and/or resistance give rise to large RC time constants, increasing the signal delays along these paths. Since the values of R' and C' are established by the processing, and are not under the control of the chip designer, control of the parasitics is achieved in the physical design and layout of the network. These effects are particularly important on long interconnects, such as those used as data buses and clock distribution networks.

3.7.4 Process Values

Basic values for the parasitic resistances and capacitances are established by the process flow. In order to make the data generally applicable to arbitrary designs, the characteristics are usually specified with regard to a convenient metric such as a

---

[1] This assumes a TEM wave along the interconnect.

square or a micron. In this section, we will present typical values for the SCNA technology. It is important to remember that these are only averaged nominal values, as processing variations are unavoidable and the value for any particular wafer will be different.

Sheet Resistance

Every physical layer on the chip exhibits parasitic resistance to current flow, so that sheet resistance values are very important in high performance design. To minimize the parasitic resistance, METAL1 and METAL2 layers are used.

TABLE 3.5 SCNA Sheet Resistance Values

LAYER	POLY	ndiff	pdiff	METAL1	METAL2
R_s [Ω/square]	22	35	75	0.05	0.03

Capacitance

Parasitic capacitances are complicated by the fact that contributions exist between every pair of conducting layers. Layer-to-substrate capacitors are the most important, since every material is patterned over the substrate, and this directly affects signal flow in the circuit. Layer-to-layer capacitances exist when different materials have an insulating oxide layer separating them and the two layers physically cross each other. In addition, patterns on the same layer exhibit capacitive coupling. Both types are summarized in Figure 3.26, and Table 3.6 provides typical values for the SCNA process. Fringing capacitance for the derived layers ndiff and pdiff have been included to account for the perimeter contribution.

TABLE 3.6 SCNA Interconnect Capacitance Values

Layer-to	POLY	ndiff	pdiff	METAL1	METAL2	Units
Substrate	0.058	0.122	0.347	0.026	0.016	fF/μm^2
POLY				0.040	0.021	fF/μm^2
METAL1					0.036	fF/μm^2
Fringe		0.451	0.210			fF/μm

When modelling the effects of parasitic capacitors, it should be noted that layer-to-substrate contributions slow the operation of the circuit, while coupling capacitance due to line-to-line interactions contribute to problems in crosstalk and noise margins. Both limit the performance of high-speed, high-density circuits.

3.7.5 Parasitics in L-Edit

L-Edit provides a simple interface for entering parasitic resistance and capacitance values. This is accessed using the Layers command in the Setup window of the Menu Bar. Choosing this command opens the dialog window shown in Figure 3.27.

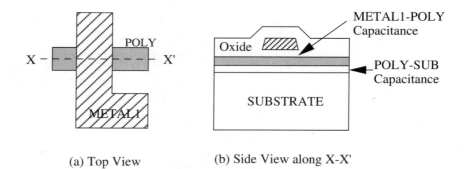

(a) Top View   (b) Side View along X-X'

Figure 3.26. Layer-to-layer capacitive coupling.

The current layer being edited is specified in the box. To enter a value for the capacitance in units of $aF/\mu m^2$ ($1\ aF = 10^{-18}\ F$), point to the capacitance box, click, and then type in the values using the keyboard; the sheet resistance is entered in the same manner. Other layers are selected by a point and click to the appropriate box in the Technology Palette. Derived layers, such as ndiff and pdiff, may not be in the displayed portion, so you may have to scroll the Technology Palette window to find the appropriate layers.

3.8 Latch-Up

Latch-up is a condition that may occur in CMOS integrated circuits where

- The circuits cease to operate;
- There is excessive consumption of current from the power supply, which may cause overheating and chip failure; and,
- The only way to take the chip out of latch-up is to disconnect the power supply.

Latch-up originates from the n and p layers used to create nMOS and pMOS transistors in the CMOS fabrication process. As such, it can be prevented by following certain rules in the layout and electrical connections.

Let us first examine why latch-up occurs. Figure 3.28 shows a basic cross-section of an n-well CMOS chip. Note that the power supply and ground connections have been included. Following the connections from V_{DD} to ground shows the existence of a 4-layer pnpn structure:

p^+ connected to VDD;

n-well;

p-substrate;

n^+ connected to ground.

If the chip goes into latch-up, current flows from the power supply to ground as shown. The circuits do not function since they do not receive any current.

In power electronics, the pnpn layering scheme is used to create a device known as a silicon-controlled rectifier (SCR), which is used as a switched rectifier. The I-V-

---

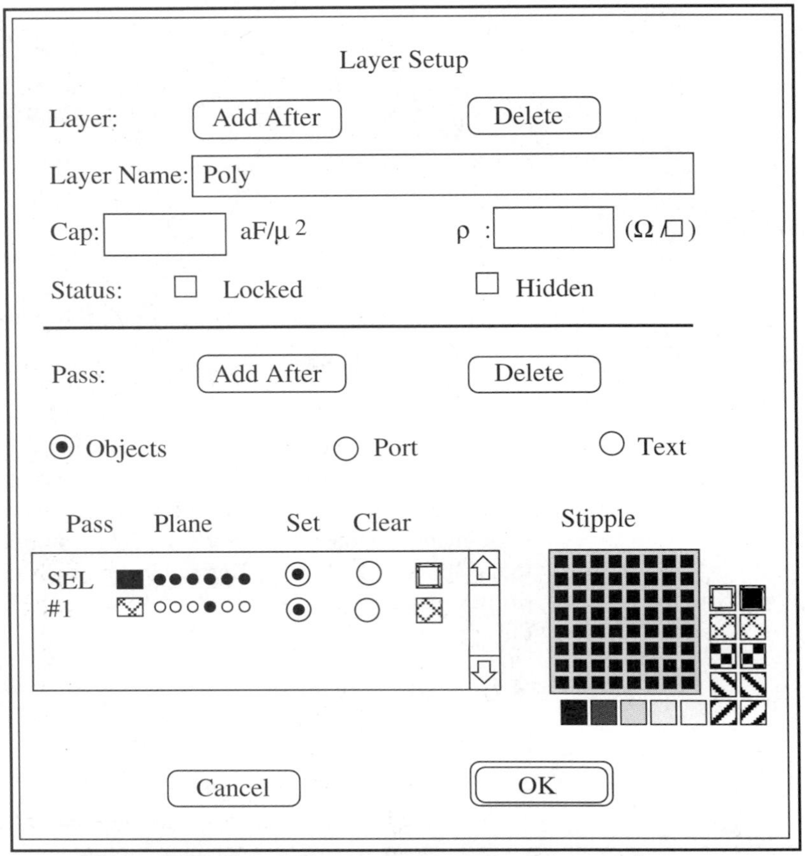

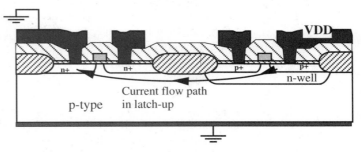

Figure 3.27. Parasitic capacitance and resistance values.

characteristics are shown in Figure 3.29. For small voltages, the structure acts like a reverse-biased pn junction, and only leakage current flows. However, if the applied voltage reaches the **break-over voltage** V_{BO}, then the curve exhibits a negative slope. This results in a very quick drop in the voltage accompanied by an increase

Figure 3.28. Origin of the latch-up problem in CMOS.

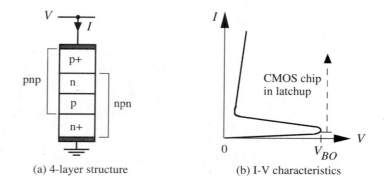

| (a) 4-layer structure | (b) I-V characteristics |

Figure 3.29. Four-layer (pnpn) device characteristics.

in the current. In CMOS, this corresponds to the chip going into latch-up.

Latch-up is often explained using bipolar transistor models. The drawing shows that both pnp and npn transistors can be visualized from the layering. Since the two transistors share the internal n and p layers, they are automatically connected in a feedback loop. Using this viewpoint, latch-up occurs when the sum of the common-base current gains equals unity:

$$\alpha_{npn} + \alpha_{pnp} = 1. \tag{3.13}$$

One approach to reducing the occurrence of latchup is to ensure that the gains of the bipolar transistors are kept small by reducing the efficiency of the emitter and base regions.

At the physical design level, latch-up prevention is achieved by adhering to a set of rules that are designed to either distribute the voltages throughout the layered regions, or reduce the current gain of the bipolar transistors. The following items are common to most processes.

- Use **guard rings** around MOSFETs;
- Provide liberal substrate contacts to ground, and n-well contacts to V_{DD};
- Obey all design rules.

Guard rings are p^+ regions connected to ground surrounding nMOSFETs, or n^+ regions connected to V_{DD} surrounding pMOSFETs, that are added to reduce the gains of the parasitic bipolar transistors.

Latch-up prevention techniques are usually specified for a given fabrication process, and should be followed to ensure a functional design.

3.9 References

R3.1 H.B. Bakoglu, Circuits, **Interconnections, and Packaging for VLSI**, Addison-Wesley, Reading, MA, 1990. An excellent examination of many issues involved in physical design of integrated circuits.

R3.2 A.K. Goel, High-Speed VLSI Interconnections, John Wiley & Sons., New York, 1994.

R3.3 B. Spinks, **Introduction to Integrated Circuit Layout**, Prentice-Hall, Englewood Cliffs, NJ, 1985. Although this book is about the layout of nMOS integrated circuits, it provides a good introduction to general principles.

R3.4 R.R. Troutman, **Latchup in CMOS Technology**, Kluwer Academic Publishers, Norwell, MA, 1986.

R3.5 J.P. Uyemura, **Fundamentals of MOS Digital Integrated Circuits**, Addison-Wesley, Reading, MA, 1988.

3.10 Exercises

E3.1 Consider two parallel n^+ lines as shown in Figure 3.4. Assume that the doping densities are $N_d=10^{19}$ cm^{-3} and $N_a=10^{15}$ cm^{-3}.

(a) Find the built-in voltage V_{bi}.

(b) Calculate the value x_p of a p-side depletion region when a reverse bias of 5 volts is applied.

E3.2 Use L-Edit to create the following patterns, and then run a DRC to check the layout. Correct any design rule violations that you find.

(a) Two parallel POLY lines that are $2\mu m$ wide and spaced $1\mu m$ apart.

(b) Two parallel POLY lines that are $1\mu m$ wide and spaced $2\mu m$ apart.

(c) Two parallel POLY lines that are $2\mu m$ wide and spaced $2\mu m$ apart.

(d) Two parallel METAL lines that are $2\mu m$ wide and spaced $3\mu m$ apart.

(e) A $2\mu m$ wide POLY line that is parallel to a $3\mu m$ wide METAL1 line, with the two lines spaced $2\mu m$ apart.

E3.3 Use the values in Tables 3.5 and 3.6 to compute the values of R_{line} and C_{line} for interconnects made as follows.

(a) A POLY line that is $2\mu m$ wide and 41 μm long.

(b) A METAL line that is $3\mu m$ wide and 70 μm long.

Chapter 4

MOSFETs

In this chapter, we will review some basic principles of MOSFET theory and operation. Simple device equations will be introduced to provide a basis for modelling. Then we will examine the interplay between the layout and the transistor characteristics.

4.1 General Observations

MOSFETs are the fundamental devices used in a CMOS integrated circuit. In fact, most logic networks are constructed entirely of MOSFETs. Other electrical elements, such as resistors and capacitors, appear only as parasitics from the physical implementation. This chapter examines the operation of MOSFETs and how the layout geometry affects the electrical characteristics. This will set the foundation for the discussion of switching circuits in the next chapter.

As discussed in Chapter 2, MOSFETs are the primary switching devices in high-density IC design because of their small size and the ability to interchange drain and source. When performing CMOS circuit design, however, most of the electrical properties of a MOSFET are already set by the processing. Circuit design revolves around setting the dimensions of the transistor since

- The dimensions of a MOSFET determine the electrical characteristics of the device.

Layout and circuit design are thus merged into a single discipline, and the performance of a circuit is directly related to the electrical parameters.

4.2 Basic Structure

The current-voltage (*I-V*) equations of a MOSFET give us the ability to calculate circuit parameters, which in turn provides the basis for design. In this section, we examine the structure and operation of the transistor from a quantitative viewpoint.

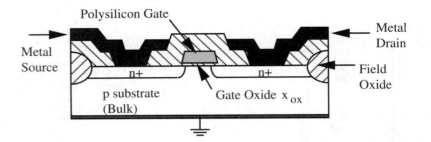

Figure 4.1. Structure of an n-channel MOSFET.

The physical structure of an n-channel MOSFET (which we will often refer to as nFET or nMOS) is shown in Figure 4.1. The device has three primary terminals labeled as the gate, drain, and source. The p-type bulk constitutes the fourth electrode; as implied in the drawing, the bulk is always assumed to be electrically connected to ground. A p-channel MOSFET (pFET or pMOS) has the same structure, but the doping polarities are reversed, i.e., n-regions are changed to p-regions, and vice-versa. As discussed in the previous chapter, the field oxide surrounds the MOSFET to ensure that the device is electrically isolated from its neighbors.

The most important geometrical parameters in the MOSFET are the gate oxide thickness, x_{ox}; the channel length, L; and the channel width, W. The value of x_{ox} is set in the fabrication, so that the circuit designer has no control over its value. The channel length and width, on the other hand, are determined by the layout of the transistor, and are the primary design variables used in circuit design. Figure 4.2 provides a perspective view of the MOSFET to more clearly understand the meaning of the channel dimensions L and W.

4.2.1 Operation

The value of the drain current I_D flowing through a MOSFET is determined by the values of V_{GS} (the gate-to-source voltage) and V_{DS} (the drain-to-source voltage) applied to the transistor. A source-to-bulk voltage (or **body bias** voltage) V_{SB} also affects the current flow characteristics in a lesser manner.

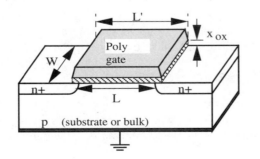

Figure 4.2.
Perspective view of an
n-channel MOSFET.

To understand the operational modes of a MOSFET, we note that the gate-oxide-semiconductor layers form a basic capacitor structure where the silicon dioxide[1] acts as the insulator between two conducting "plates." When discussing MOS devices, the capacitance of this structure is described by C_{ox}, and is termed the oxide capacitance per unit area with units of *farads per cm²* as calculated from

$$C_{ox} = \frac{\varepsilon_{ox}}{x_{ox}}.$$

(4. 1)

In this equation, $\varepsilon_{ox} = (3.9)(8.854 \times 10^{-14}) \approx 3.45 \times 10^{-13}$ *farads/cm* is the permittivity of silicon dioxide and x_{ox} is the thickness of the oxide in centimeters. As an example, suppose that $x_{ox} = 200 \times 10^{-8}$ *cm* (200 Å). This gives an oxide capacitance of

$$C_{ox} \approx 1.73 \times 10^{-8}$$

(4. 2)

in units of *farads/cm²*. A more convenient length metric is the micrometer (μm), also called a *micron*, with $1 \ \mu m = 10^{-6} \ m$. Then, the value is

$$C_{ox} \approx 0.173 fF / \mu m^2 \ ,$$

(4. 3)

where *fF* denotes a femtofarad; 1 *fF* is equal to 10^{-15} farads. The total gate capacitance C_G in units of farads is calculated using

$$C_G = C_{ox} WL' ,$$

(4. 4)

where W is the channel width, and L' is the **drawn channel length** (the length of the poly), both in units of centimeters. Note that both L and L' are different quantities, as shown in Figure 4.2; in particular, $L'>L$.

In the simplest viewpoint, current flow in the n-channel MOSFET is controlled by the gate-source voltage V_{GS}. For small values of V_{GS}, the n⁺ drain and source regions are separated by the p-type substrate, making it difficult for current flow to exist between the two terminals. In this case, $I_D \approx 0$, and the transistor is said to be in **cutoff**. If V_{GS} is increased to a value $V_{GS}>V_{Tn}$, where V_{Tn} is called the **threshold voltage**, then the gate capacitor structure induces a thin layer of electrons at the surface of the silicon. This provides a channel between the source and drain, establishing current flow. When the MOSFET is biased in this manner, the device is said to be **active**. The value of I_D is set by V_{GS} and V_{DS} (the drain-to source voltage).

We note at this point that the threshold voltage V_{Tn} is established in the fabrication sequence, and is usually specified by a nominal value for use in circuit design. It is a sensitive function of x_{ox}, the doping densities of the gate and substrate, and the physical properties of the materials, but can be controlled to some extent using ion implantation.

---

[1] Silicon dioxide is quartz glass.

4.2.2 Square-Law Model

The electrical characteristics of a MOSFET can be approximated using a simplified square law relation between the drain current I_D and the voltages V_{DS} and V_{GS}; these are shown in Figure 4.3 for an n-channel MOSFET. The source-to-bulk voltage V_{SB} affects the current flow through the transistor by altering the threshold voltage V_{Tn}.

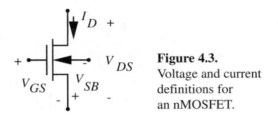

Figure 4.3.
Voltage and current definitions for an nMOSFET.

There are two distinct operational regions for an active MOSFET: **non-saturation** (or linear) and **saturation**. The mode depends on the relative values of V_{DS} and V_{GS} through the saturation voltage $V_{DS,sat}$ as defined by

$$V_{DS, sat} = (V_{GS} - V_{Tn}).$$

(4. 5)

For small drain-source voltages that satisfy $V_{DS} \leq V_{DS, sat}$, the MOSFET is said to be non-saturated with

$$I_D = \frac{\beta_n}{2} [2 (V_{GS} - V_{Tn}) V_{DS} - V_{DS}^2],$$

(4. 6)

where

$$\beta_n = k'_n \left(\frac{W}{L}\right)$$

(4. 7)

is called the **device transconductance** with units of *amps/volts²*. k'_n is called the **process transconductance** (also with units of *amps/volts²*), and is given by

$$k'_n = \mu_n C_{ox},$$

(4. 8)

with μ_n as the electron surface mobility. In a typical CMOS process, $\mu_n \approx 550\text{-}600$ *cm²/V-sec*. The factor *(W/L)* is a dimensionless geometrical layout parameter known as the **aspect ratio**, and is simply the **channel width** *W* divided by the **channel length** *L*. The central problem in the electrical design of CMOS circuits often reduces to finding the aspect ratios that give the desired performance. The smallness of MOSFETs is evident from the fact that *L* is smaller than the minimum value specified by the design rule set.

If a large drain-source voltage is applied such that $V_{DS} \geq V_{DS, sat}$, then the MOSFET is operating in **saturation** such that the drain current is described by

$$I_D = \frac{\beta_n}{2} (V_{GS} - V_{Tn})^2 [1 + \lambda (V_{DS} - V_{DS,sat})] \tag{4.9}$$

where λ is the **channel-length modulation factor** with units of *volts^{-1}*. Figure 4.4 shows a typical family of curves for a MOSFET. Each line corresponds to the behavior with a particular value of gate-source voltage V_{GS}. The border between saturation and non-saturation is given by the saturation current

$$I_{sat} = \frac{\beta_n}{2} V_{DS,sat}^2 . \tag{4.10}$$

The increase of the drain current in a saturated MOSFET is governed by the value of λ. In digital design, it is common to approximate $\lambda=0$ to simplify hand calculations. In this case, the current in a saturated transistor has a value

$$I_D \approx \left(\frac{\beta_n}{2}\right) (V_{GS} - V_{Tn})^2 . \tag{4.11}$$

Analog circuits, on the other hand, are often sensitive to the value of λ, so this parameter is usually specified for those designs.

4.2.3 Body Bias

The threshold voltage V_{Tn} is affected by the source-bulk voltage V_{SB} according to

$$V_T = V_{T0} + \gamma (\sqrt{2|\phi_F| + V_{SB}} - \sqrt{2|\phi_F|}) , \tag{4.12}$$

where V_{T0} is the **zero body-bias threshold voltage** (when $V_{SB}=0$), γ is the body-bias factor

$$\gamma = \frac{\sqrt{2q\varepsilon_{Si}N_a}}{C_{ox}} \tag{4.13}$$

with units of $[V^{1/2}]$, and $|\phi_F|$ is the bulk Fermi potential

$$|\phi_F| = \left(\frac{kT}{q}\right) ln\left(\frac{N_a}{n_i}\right) . \tag{4.14}$$

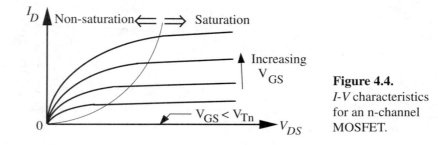

Figure 4.4.
I-V characteristics for an n-channel MOSFET.

The parameter N_a in both equations is the acceptor (boron) doping density in the p-type substrate; typically, N_a is on the order of 10^{15} cm^{-3}. This equation shows that the threshold voltage increases with increasing body bias voltage V_{SB}, making it more difficult to turn on. Figure 4.5 shows the general behavior.

4.2.4 Process Parameters

The basic electrical parameters of a MOSFET are set by the processing. For an n-channel transistor, these are summarized as follows:

- V_{T0n}, the zero-body bias threshold voltage. This is a positive number, typically between about 0.50 and 1.00 volts in current CMOS processing.

- k'_n, the process transconductance $[A/V^2]$. The process transconductance is determined by the value of C_{ox}, which is inversely proportional to the oxide thickness x_{ox}. A typical range for current CMOS processing is 50-200 $[\mu A/V^2]$.

- γ, the body-bias parameter $[V^{1/2}]$. This varies with substrate doping and oxide capacitance.

- λ, the channel-length modulation factor $[V^{-1}]$, an empirical factor.

It is important to remember that the circuit designer cannot alter these values. Moreover, processing variations usually require that we work with nominal values (not exact numbers) when designing circuits. This complicates the designer's job, since it is more difficult to design a circuit that will still operate within reasonable limits.

4.3 MOSFET Parasitics

CMOS logic design is simplified by the fact that MOSFETs can be modelled as voltage-controlled switches. This is due to the distinct modes of cutoff and active operation, depending on the value of the gate-source voltage V_{GS} relative to the threshold voltage V_T. Although the switching behavior is straightforward, the electrical characteristics of the MOSFET are affected by the parasitic capacitances and resistances inherent in the device.

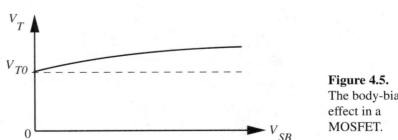

Figure 4.5.
The body-bias effect in a MOSFET.

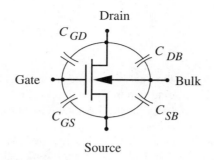

Drain

C_{GD} C_{DB}

Gate   Bulk

C_{GS} C_{SB}

Source

Figure 4.6.
Parasitic capacitance
model for a MOSFET.

4.3.1 Parasitic Capacitances

Figure 4.6 illustrates the primary capacitances associated with a MOSFET. These may be classified as gate or depletion, depending on their origin.

Switching action in a MOSFET is due to the MOS capacitance per unit area C_{ox}. When looking into the gate of a MOSFET, we often use

- C_G: The gate capacitance,

as the total input capacitance. This can be calculated from

$$C_G = C_{ox}WL' ,$$
(4. 15)

where L' is the drawn channel length (as measured by the poly gate geometry). It is a reasonable approximation for the capacitance between the gate and the bulk electrodes.

The parasitic coupling effects between the gate and the source or drain electrodes are modelled by introducing

- C_{GS}: The gate-to-source capacitance and
- C_{GD}: The gate-to-drain capacitance,

for use in the circuit analysis. In general, both are functions of the voltages V_{GS} and V_{DS}. A coarse (but invitingly simple) estimate ignores the voltages to write

$$C_{GS} \approx \frac{1}{2}C_G \approx C_{DS} .$$
(4. 16)

Although this is, at best, a first-order approximation, it does provide reasonable values for initial design calculations. More accurate values are discussed in [1].

Depletion capacitance is due to the change in dopant polarity at a pn junction; the source-bulk and drain-bulk capacitances C_{SB} and C_{DB} originate from this effect. These are functions of the reverse-bias voltage across the junctions. In the general case, the zero-bias junction capacitance per cm^2 is usually denoted by C_{j0} such that the total capacitance as a function of the reverse voltage V_r is obtained from the expression

$$C = \frac{C_{j0}A}{\left(1 + \dfrac{V_r}{\phi_o}\right)^m}. \qquad\qquad \textbf{(4. 17)}$$

In this equation, A is the total junction area and ϕ_O is the built-in voltage (or potential barrier). The grading parameter m is used to describe the doping profile; two special cases are $m=1/2$ (a step profile junction), and $m=1/3$ (a linearly graded junction).

Applying this formula to a MOSFET requires that we examine the three-dimensional geometry more carefully. Consider the n^+ region embedded in a p-type substrate as shown in Figure 4.7. The acceptor doping $N_{a,sw}$ on the sidewall regions is usually higher than the doping N_a on the bottom portion due to additional ion implantation steps that are used to control the threshold voltage of the transistor and the field regions. The structure is thus divided into "bottom" and "sidewall" sections, which results in two distinct values of zero-bias capacitance values. These are denoted by C_{j0} F/cm^2 and $C_{j,sw}$ F/cm for the bottom and sidewall regions, respectively. The zero-bias depletion capacitance is then calculated from

$$C_D = C_{j0}A + C_{j,sw}P, \qquad\qquad \textbf{(4. 18)}$$

where A is the area of the bottom, and P is the length of the perimeter. In terms of the geometry shown in Figure 4.8, we see that $A \approx WX$ and $P \approx 2(W+X)$, ignoring lateral doping effects. Note that increasing the channel width W increases both the bottom and sidewall contributions to the depletion capacitance.

For hand calculations it is easiest to use C_{j0} in units of $fF/\mu m^2$ and $C_{j,sw}$ in units of $fF/\mu m$; then A is specified in μm^2 while P has units of μm and values can be read directly from the layout. Remember that the depletion capacitance exists at both the drain and source sides of a MOSFET. Both need to be included when estimating parasitics.

4.3.2 Drain-Source Resistance

The drain-to-source resistance of a MOSFET is due to the resistivity of the channel. If we attempt to define a linear resistance R for the transistor using

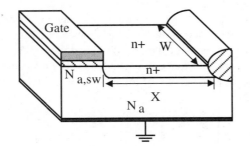

Figure 4.7.
Drain/source geometry
for a MOSFET.

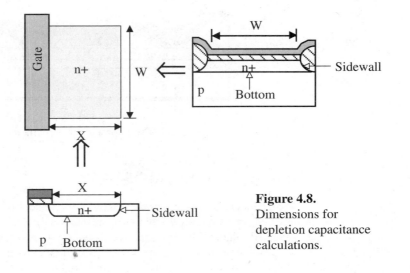

Figure 4.8.
Dimensions for
depletion capacitance
calculations.

$$R = \frac{V_{DS}}{I_D},$$ (4. 19)

then the nonlinear features of the device are immediately obvious when trying to determine which equation is appropriate for the drain current. To simplify matters, we can estimate the *effective* drain-source resistance of an nFET by using the linear, time-invariant value obtained from

$$R_n \approx \frac{1}{k'_n \left(\dfrac{W}{L}\right)(V_{DD} - V_{Tn})}.$$ (4. 20)

Although this is reasonable as a first-order approximation, it should never be used to calculate the circuit performance in critical situations. Computer simulations are required in this case. The equation is useful for comparing alternate circuit designs.

4.3.3 Other Parasitic Resistances

In a complete model, two other MOSFET resistances should be included. The first is R_{n+}, which is due to the finite conductivity of the n^+ drain and source regions. These contributions are in series with the effective drain-to-source resistance R_n.

The remaining parasitic is the contact resistance R_c at the metal-semiconductor junctions. The use of n^+ drain and source regions reduces this parasitic to a reasonably small value by creating a tunnel-ohmic contact. It is often ignored in the analysis unless the device falls in a critical chain of circuit where the exact timing is important.

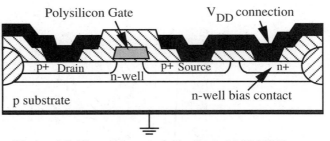

Figure 4.9. The geometry of a p-channel MOSFET.

4.4 p-Channel MOSFETs

A p-channel MOSFET is the electrical complement of the n-channel MOSFET. In essence, the structure of the two are identical, with only the n and p polarities reversed. Figure 4.9 shows the cross-section of a p-channel MOSFET in an n-well technology (as discussed in Chapter 2). Note that the n-well region acts as the bulk electrode for the pFET, and is electrically connected to the power supply voltage V_{DD} as shown. The geometrical parameters are again the channel width W and the channel length L, such that the aspect ratio (W/L) controls the current flow. It should be noted that the device transconductance is

$$k'_p = \mu_p C_{ox},\qquad(4.21)$$

where μ_p is the surface hole mobility and C_{ox} is the oxide capacitance per unit area. The value of C_{ox} is the same as that used for nMOSFETs on the chip. However, the hole mobility is less than the electron mobility, and

$$k'_p < k'_n\qquad(4.22)$$

always holds[1]; $k'_n \approx 2\,k'_p$ is a reasonable scaling between the two in CMOS. Since the device transconductance assumes the form

$$\beta_p = k_p \left(\frac{W}{L}\right)_p,\qquad(4.23)$$

equal size pFET and nFET devices will have $\beta_p < \beta_n$.

Current flow through a pFET is controlled by the source-gate voltage V_{SG} and the source-drain voltage V_{SD} as shown in Figure 4.10. The threshold voltage of a pFET is usually referenced to the gate-to-source voltage V_{GS}, and is stated as a negative number $V_{Tp} < 0$ in the process specifications. However, it is much simpler to describe the conduction using the source-gate voltage V_{SG} and the absolute value of the threshold voltage $|V_{Tp}|$.

---

[1] Note that the mobility is also reduced because the n-well doping of the pFET is larger than the p-substrate doping of an nFET.

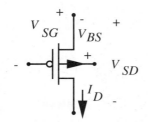

Figure 4.10.
Current and voltages
for a p-channel MOSFET.

Cutoff in a pFET occurs when $V_{SG} < |V_{Tp}|$ and gives $I_D \approx 0$. To obtain active operation where current flows, the source-gate voltage must be elevated to $V_{SG} \geq |V_{Tp}|$. The border between non-saturated and saturated current flow is determined by the value of V_{SD} relative to the saturation voltage

$$V_{SD,\,sat} = (V_{SG} - |V_{Tp}|) .$$ (4. 24)

For small source-drain voltages that satisfy $V_{SD} < V_{SD,sat}$, the pFET is non-saturated with a drain current

$$I_D = \frac{\beta_p}{2} [2 (V_{SG} - |V_{Tp}|) V_{SD} - V_{SD}^2] .$$ (4. 25)

If the source-drain voltage is large such that $V_{SD} \geq V_{SD,sat}$, then the transistor is saturated with

$$I_D = \frac{\beta_p}{2} (V_{SG} - |V_{Tp}|)^2 .$$ (4. 26)

Channel-length modulation effects can be included in the same manner as for nMOSFETs by multiplying the current by the factor $[1 + \lambda_p(V_{SD} - V_{SD,sat})]$.

The parasitic elements of a p-channel MOSFET are the same as those discussed for the nFET. The source-drain resistance can be estimated using the formula

$$R_p \approx \frac{1}{k'_p \left(\dfrac{W}{L} \right)_p (V_{DD} - |V_{Tp}|)} ,$$ (4. 27)

while the basic capacitance equations are still valid with appropriate changes in the numerical values of C_{j0} and $C_{j,sw}$.

A p-channel MOSFET has the same basic process-related parameters as an nFET:

- V_{T0p}, the threshold voltage. This is a negative number, typically between about -0.60 and -1.00 volts.

- k'_p, the process transconductance [A/V^2]. The hole mobility is less than the electron mobility, giving values for k'_p that are less than the corresponding factors for an nFET. A reasonable range for current CMOS processing is around 15-70 [$\mu A/V^2$].
- γ, the body-bias parameter [$V^{1/2}$]. This varies with the n-well doping and oxide capacitance.
- λ, the channel-length modulation factor [V^{-1}], an empirical parameter that we will ignore in simple digital circuit estimates.

Depletion capacitance values are denoted by C_j and $Cj_{,sw}$, but the values are different from those used for nFETs because the doping densities are different.

4.5 MOSFET Modelling

CMOS circuits can be viewed as either switching networks or electronic networks, depending on whether one is checking the logic or examining the performance. MOSFET models can be constructed for either case.

4.5.1 Logic Models

Switching circuits that implement Boolean logic functions can be designed by using very simple models for MOSFETs. To introduce this approach, we need to establish a relationship between voltage levels and Boolean logic 0 and logic 1 values. CMOS circuits are usually designed to operate with a single power supply voltage V_{DD} applied with respect to ground; the ground node is numerically taken to be 0 V. The exact value of V_{DD} varies with the process, but common choices are 5V, 3.3V, 3.0V, and smaller. Because of the use of a single positive power supply, digital CMOS is described using a **positive logic convention**, where, ideally,

> **logic 0:** 0 volts

> **logic 1:** V_{DD}

defines the relationship between Boolean variables and voltages. In realistic circuits, the discrete voltages are replaced by a range of values. Qualitatively, we simply write that

> **logic 0:** Low voltages (close to 0) and

> **logic 1:** High voltages (close to V_{DD}),

with the actual values defined by the operation of the circuit. This will be discussed in the next chapter in the context of static logic gate design. For the moment, it is sufficient to use the ideal definition of logic 0 and logic 1 voltages.

At the simplest level of modelling, a MOSFET may be viewed as a voltage-controlled switch. Since nMOSFETs and pMOSFETs are electrical opposites of one another, they act in a complementary manner when used as switches.

Consider first the n-channel MOSFET (nFET or nMOS). Ideally, this acts as the

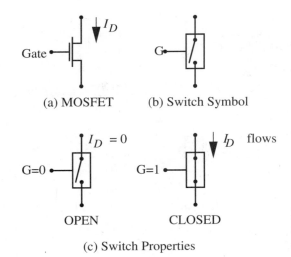

(a) MOSFET   (b) Switch Symbol

OPEN   CLOSED

(c) Switch Properties

Figure 4.11.
Switching model for
an n-channel MOSFET

switch shown in Figure 4.11; note that the drawing also provides the schematic symbol for the transistor. The operation is straightforward. With a logic 0 applied to the gate, the switch is OPEN, so that there is no drain current flow: $I_D = 0$. On the other hand, applying a logic 1 to the gate closes the switch; a CLOSED switch allows current to flow.

A p-channel MOSFET (pFET or pMOS) is exactly opposite. A logic 0 applied to the gate gives a CLOSED switch, and current flows between source and drain. If a logic 1 is applied to the gate, the switch is OPEN, and the current is zero. Figure 4.12 summarizes the pFET switching properties. Note that the pFET switch symbol uses an *inversion bubble* at the input; this allows us to distinguish between the two types of switches, and also serves as a useful reminder that the two act in opposite manners.

Caution must be exercised when using these simplified switching models, since they ignore many important electrical characteristics of MOSFETs such as the turn-on voltage, drain-to-source resistance, and parasitic capacitance. However, they are quite useful for the analysis and design of basic logic networks, since it is the topology of the circuit that determines the logic function. This allows us to structure the logic by proper placement of the transistors; optimizing the performance, on the other hand, requires more complex models.

4.5.2 Electrical Models

Logic models view MOSFETs as simple voltage-controlled switches. Although this is sufficient for describing the logic functions, it ignores the electrical characteristics of the transistors. In circuit applications, it is important to include the parasitic elements so that an accurate analysis can be performed. There are two distinct signal paths through a MOSFET: one looks into the gate, while the other is the current flow between the drain and the source. It is straightforward to develop simple

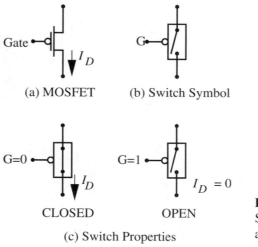

(a) MOSFET   (b) Switch Symbol

G=0   CLOSED   G=1   OPEN $I_D = 0$

(c) Switch Properties

Figure 4.12.
Switching model for
a p-channel MOSFET

switching models for each case to help visualize the device operation in a digital network.

Gate Input

This situation is shown in Figure 4.13. For the purpose of computing the load to the circuit, the MOSFET is modelled as a simple capacitor with a value of the gate capacitance C_G. This can be further split into equal gate-source and gate-drain contributions to estimate the coupling effects.

Drain-Source Current Flow

When used in a switching circuit, the drain-source characteristics can be modelled as a voltage controlled switch with parasitic resistance and capacitance as shown in Figure 4.14. This is particularly useful for estimating transient response times.

The parasitic elements originate from the characteristics of the channel and the intrinsic capacitances of the MOSFET structure. Consider first the nFET. For gate-

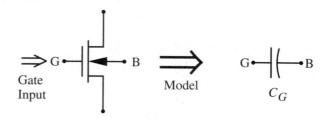

Gate Input   Model C_G

Figure 4.13. Input gate capacitance.

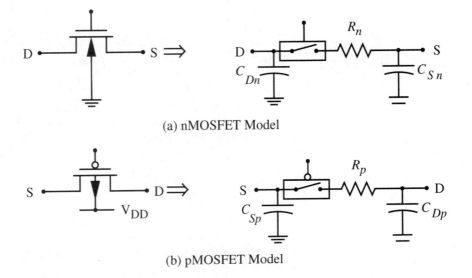

(a) nMOSFET Model

(b) pMOSFET Model

Figure 4.14. Drain-source switch model.

source voltages $V_{GS}<V_T$, the n-channel MOSFET is in cutoff. Since the drain current is zero, this situation is modelled as an open switch. If the gate-source voltage is large such that $V_{GS}>V_T$, then the device is active and the switch is closed. Current flow is then affected by the parasitic resistance R_n and capacitance at both sides of the transistor. In general, the model works best if C_{Dn} and C_{Sn} include both pn junction and gate-channel contributions. The switching behavior of a pFET is exactly opposite; the switch is closed with a logic 0 (low) voltage applied to the gate, while a large gate voltage (logic 1) gives an open circuit.

4.5.3 The MOSFET RC Time Constant

In circuit design, the aspect ratio (*W/L*) becomes the primary design variable for adjusting the electrical performance. Since the drain current I_D is proportional to β, this has the effect of adjusting the current flow through a particular transistor. However, it is important to understand how the parasitics change. This can be described by using the concept of a time constant.

A MOSFET exhibits both drawn-source resistance and parasitic capacitance at both the source and drain. Let us define a device time constant $\tau = RC$ that applies to one side of the transistor (to the source or drain side of the device) by multiplying the resistance and the appropriate capacitance together. For example, on the source side of an nMOSFET, we have

$$\tau = R_n (C_{GS} + C_{SB}).$$ (4. 28)

This includes both the gate-source and source-bulk capacitances. Using the equa-

---

tions discussed above, this can be written as

$$\tau = \frac{\frac{1}{2} C_{ox} WL' + C_j WX + C_{jsw} 2 (W + X)}{k_n \left(\dfrac{W}{L}\right) (V_{DD} - V_{Tn})}, \qquad \text{(4. 29)}$$

where the dimensions of the source n^+ region have been assumed to be $(W \times X)$.

Now suppose that the width of the transistor is increased from W to a new value $W' > W$. The new time constant is

$$\tau' = \frac{\frac{1}{2} C_{ox} W'L' + C_j W'X + C_{jsw} 2 (W' + X)}{k_n \left(\dfrac{W'}{L}\right) (V_{DD} - V_{Tn})}. \qquad \text{(4. 30)}$$

Comparing this with the original equation for τ shows that, were it not for the last term in the numerator, the expressions would be the same. Because of this, we are sometimes able to estimate

$$\tau' \approx \tau, \qquad \text{(4. 31)}$$

in other words, the internal time constant is approximately constant as the channel width W is varied. If, on the other hand, the sidewall term that is proportional to X is not small compared to W, then $\tau' < \tau$. These observations are useful when trying to optimize the switching speed of a digital circuit.

4.6 MOSFET Layout

The aspect ratio *(W/L)* is the layout parameter that determines the maximum current flow through a MOSFET. Although the dimensions of a transistor are arbitrary, a high-speed design requires large current flow levels, so that L is usually (but not always!) taken to be the minimum value permitted by the process lithography. The **channel width** then defaults to being the primary design parameter, although we usually specify the aspect ratio *(W/L)* since this is the factor that appears in the equations for the drain currents.

Layout of a MOSFET using L-Edit is very straightforward. An n-channel device is constructed by creating an n^+ region **ndiff** defined by

> **ndiff** = (**ACTIVE**) AND (**NSELECT**).

A **POLY** over **ndiff** creates the transistor. The drawing steps for creating the nFET are as follows.

1. Construct an **ACTIVE** box/polygon.

2. Surround **ACTIVE** with **NSELECT**. The intersection of the two is **ndiff**.

3. Create a **POLY** box that crosses completely over ndiff and extends beyond the active area. This creates the gate.

The actual drawing sequence is not important. However, all design rules should be obeyed. Figure 4.15 shows the layers used for a simple transistor.

One complication that arises is that the length L' of the POLY is larger than the actual channel length L by a factor of $2L_D$ where L_D is the **lateral diffusion length**:

$$L = L' - 2L_D. \tag{4.32}$$

The channel width W is also modified due to active area encroachment when growing the field oxide. In this case, the electrical channel width is given by

$$W = W' - 2(\Delta W) \tag{4.33}$$

where W' is the drawn width (set by the dimensions of the ACTIVE polygon), and ΔW is the reduction due to encroachment and other effects.

A p-channel MOSFET follows the same basic order, except that the n-well must be defined. The steps are:

1. Create an **NWELL** region for the pMOSFET.

2. Construct an **ACTIVE** box/polygon for the transistor.

3. Surround **ACTIVE** with **PSELECT**. The intersection of the two is **pdiff**.

4. Draw a **POLY** box over **pdiff** for the gate.

5. Provide an **ACTIVE** and **NSELECT** box within **NWELL** for the n-well contact (to V_{DD}).

Note that the n^+ contact formed in Step 5. is needed to bias the n-well to the power supply voltage.

POLY Routing Constraint

It is important to remember that CMOS processing creates self-aligned MOSFET structures. When you are routing POLY, note that

- Crossing POLY over ndiff=(ACTIVE) AND (NSELECT) **always** creates an n-channel MOSFET region under the POLY. Do not cross the layers unless there is a transistor in the design.

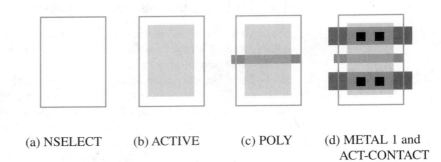

(a) NSELECT   (b) ACTIVE   (c) POLY   (d) METAL 1 and ACT-CONTACT

Figure 4.15. L-Edit MOSFET layers.

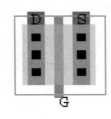
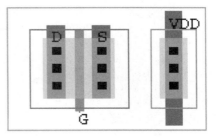

(a) nMOSFETs   (b) pMOSFET

Figure 4.16. L-Edit MOSFET layers.

Similarly,

- Crossing POLY over pdiff=(ACTIVE) AND (PSELECT) in NWELL **always** creates a p-channel MOSFET in this technology.

If interconnect is needed over ndiff or pdiff, always use METAL to avoid the formations of unwanted transistors.

L-Edit Examples

To illustrate the process, consider an n-channel MOSFET with drawn dimensions of W=14 μm and L'=2 μm. This is shown in Figure 4.16(a). A pFET with the same dimensions W=14 μm and L'=2 μm is shown in (b). Note the inclusion of the n-well contact to the power supply.

Another example is shown in Figure 4.17. This is a pair of transistors connected in parallel to yield a single MOSFET. If each transistor has an aspect ratio of (W/L), then the effective aspect ratio of the composite device is 2(W/L), doubling the current flow capacity.

4.7 The Cross-Section Viewer

The **Cross-Section Viewer** is an L-Edit tool that aids in visualizing the 3-dimensional physical structure of devices that are described by 2-dimensional layout drawings. Important concepts, such as the definition of active areas, wells, and n- and p-select, become immediately clear when examined using this feature.

To invoke the Cross-Section Viewer, simply open the window under the heading **Special** on the Menu Bar, and drag the mouse down until this option is selected as shown. Releasing the mouse button starts the option. When the Cross-Sectional Viewer is active, normal editing operations are suspended.

Operation of the Cross-Sectional Viewer is straightforward. First, you must provide L-Edit with the correct process description that translates the layouts into layers. The program will prompt you for the name of the file that contains the

Special

Generate Layers...
Clear Gen'ed Layers...
- - - - - - - - - -
DRC...
DRC Box...
Clear Error Layer...
- - - - - - - - - -
Place and Route...
- - - - - - - - - -
Extract...
- - - - - - - - - -
Cross-Section...

The Cross-Section Viewer is accessed from the Special command window.

necessary data. This information is contained in files named with a **.xst** extension. Your L-Edit disk contains extraction data for all of the technology decriptions provided with the program. After the data base has been chosen, you will be prompted to place the mouse pointer to the location that you want to view. Choose a vertical location and click the mouse button; L-Edit automatically draws a horizontal line across the layout and starts the analysis. The cross-sectional drawing will be drawn layer-by-layer in the lower half of the screen. If you want to view the buildup a single layer at a time, then choose the Single Layer option in the process prompt.

L-Edit Example

The cross-sectional view produced by L-Edit for the MOSFETs above are shown in Figure 4.18. Note the formation of the n-well in the substrate, and also the presence of the contact cuts through the oxide that connect METAL1 TO ACTIVE. The layers shown on top of the transistors represent oxide between the materials that are

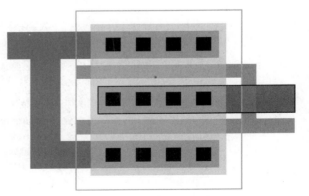

Figure 4.17. Parallel MOSFET example.

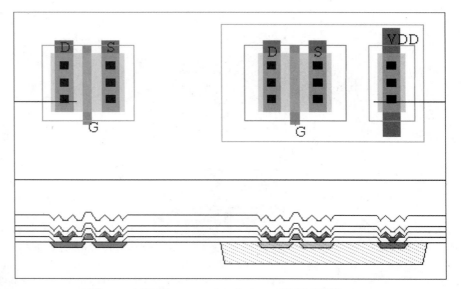

Figure 4.18. Cross sectional view of MOSFETs.

not shown, such as METAL2.

4.8 SPICE Models

Modelling MOSFETs in SPICE requires two separate statements. In general, the device connections are specified by an element listing of the form

> **M1 ND NG NS NB MODELNAME [<L=value> <W=value>**
> **+<AD=value> <PD=value> <AS=value> <PS=value>]**

where

M1 is the device name

ND, **NG**, **NS**, **NB** are the node labels for the drain, gate, source, and bulk

MODELNAME is the name of the .MODEL description

L is the channel length in meters

W is the channel width in meters

AD (**AS**) is the drain (source) area in square meters

PD (**PS**) is the drain (source) perimeter length in meters

describes specific MOSFETs in a circuit. The geometrical parameters L and W are critical to all current flow calculations, and should be specified. The drain and source dimensions given by AS, AD, PD, and PS are used to calculate parasitics.

The technology parameters from the processing are included in the .MODEL line, which has the form

.MODEL MODELNAME NMOS [Listing of parameters]

for an n-channel FET; a pFET is indicated by replacing "NMOS" by "PMOS". The parameter set includes the physical parameters that characterize the process. Note in particular, the following quantities:

- **TOX** is the gate oxide thickness (x_{ox} in our notation) in units of meters;
- **KP** represents the process transconductance k' in units of $[A/V^2]$;
- **VTO** is the zero body-bias voltage;
- **GAMMA** is the body-bias coefficient;
- **PHI** is the bulk Fermi potential $2|\phi_F|$;
- **NSUB** is the substrate doping for the MOSFET;
- **LD** is the lateral diffusion length, L_o in the current notation;
- **CJ** is the zero-bias bottom depletion capacitance in units of $[F/m^2]$;
- **CJSW** is the zero-bias sidewall capacitance in units of $[F/m]$;
- **MJ** is the bottom grading coefficient;
- **MJSW** is the sidewall grading coefficient;
- **PB** is the junction "potential barrier", i.e., the built-in potential ϕ_o;
- **LAMBDA** is the channel-length modulation factor.

It is possible to run SPICE using a minimal set of parameters consisting of VTO and KP for a very quick evaluation of the circuit operation. Providing a full set of values gives increased precision.

Several different MOSFET equation sets can be used in SPICE. In general, the equations that are used to compute the current are defined in the .MODEL line using the designation **LEVEL=N**, where N is an integer. The simplest device model is contained in the LEVEL 1 model, which is based on the simple square law equations discussed previously in the chapter. Although square law equations are useful for a qualitative understanding of MOSFETs and the simulation times are short, the results are not very accurate, particularly for short-channel devices (less than about 5 microns). LEVEL 2 modelling provides more accurate current flow equations, and also the dominant small geometry effects. More precise simulations are based on LEVEL 3, BSIM, and higher models. The models that are available to you depend on the version and maker of the SPICE you are using, so you should check the operating manual to see what options are available.

A typical set of MOSIS SCNA .MODEL LEVEL 2 statements are shown below for nMOS and pMOS devices.

This LEVEL 2 MOSFET listing is provided on the L-Edit disk in the file scna.spc

```
.MODEL CMOSN NMOS LEVEL=2 LD=0.250000U TOX=417.000008E-10
+ NSUB=6.108619E+14 VTO=0.825008 KP=4.919000E-05 GAMMA=0.172
+ PHI=0.6 UO=594 UEXP=6.682275E-02 UCRIT=5000
+ DELTA=5.08308 VMAX=65547.3 XJ=0.250000U LAMBDA=6.636197E-03
+ NFS=1.98E+11 NEFF=1 NSS=1.000000E+10 TPG=1.000000
+ RSH=32.740000 CGDO=3.105345E-10 CGSO=3.105345E-10
+ CGBO=3.848530E-10 CJ=9.494900E-05 MJ=0.847099 CJSW=4.410100E-10
```

+ MJSW=0.334060 PB=0.800000
* Weff = Wdrawn - Delta_W
* The suggested Delta_W is -0.25 um

.MODEL CMOSP PMOS LEVEL=2 LD=0.227236U TOX=417.000008E-10
+ NSUB=1.056124E+16 VTO=-0.937048 KP=1.731000E-05 GAMMA=0.715
+ PHI=0.6 UO=209 UEXP=0.233831 UCRIT=47509.9
+ DELTA=1.07179 VMAX=100000 XJ=0.250000U LAMBDA=4.391428E-02
+ NFS=3.27E+11 NEFF=1.001 NSS=1.000000E+10 TPG=-1.000000
+ RSH=72.960000 CGDO=2.822585E-10 CGSO=2.822585E-10
+ CGBO=5.292375E-10 CJ=3.224200E-04 MJ=0.584956 CJSW=2.979100E-10
+ MJSW=0.310807 PB=0.800000
* Weff = Wdrawn - Delta_W
* The suggested Delta_W is -1.14 um

The LEVEL 2 parameters are reasonable for a 2-micron geometry.

Circuits that use smaller devices should be simulated using LEVEL 3, BSIM, or an appropriate parameter set that accounts for small device effects. A sample set of LEVEL 3 parameters for the mhp_n12 technology are provided below.

.MODEL CMOSN NMOS LEVEL=3 PHI=0.600000 TOX=2.0500E-08
+ XJ=0.200000U TPG=1 VTO=0.7802 DELTA=1.6820E+00 LD=1.5380E-07
+ KP=1.0183E-04 UO=604.5 THETA=9.9190E-02 RSH=6.0900E+01
+ GAMMA=0.4692 NSUB=1.8820E+16 NFS=1.980E+12
+ VMAX=1.7930E+05 ETA=6.2960E-02 KAPPA=8.6860E-02
+CGDO=3.8861E-10 CGSO=3.8861E-10 CGBO=3.7701E-10 CJ=3.1603E-04
+MJ=1.0713 CJSW=1.3284E-10 MJSW=0.119521 PB=0.800000
* Weff = Wdrawn - Delta_W
* The suggested Delta_W is 4.6920E-07

.MODEL CMOSP PMOS LEVEL=3 PHI=0.600000 TOX=2.0500E-08
+ XJ=0.200000U TPG=-1 VTO=-0.8740 DELTA=2.1510E+00 LD=5.1830E-08
+ KP=3.0741E-05 UO=182.5 THETA=1.2750E-01 RSH=9.0000E+01
+ GAMMA=0.3806 NSUB=1.2380E+16 NFS=3.460E+12
+VMAX=3.3530E+05 ETA=1.5890E-01 KAPPA=9.8690E+00
+ CGDO=1.3096E-10 CGSO=1.3096E-10 CGBO=3.5622E-10 CJ=4.7512E-04
+ MJ=0.5078 CJSW=1.6127E-10 MJSW=0.225871 PB=0.850000
* Weff = Wdrawn - Delta_W
* The suggested Delta_W is 4.0460E-07

BSIM parameters are a bit more mysterious to decipher, and have not been included here. Also, it should be mentioned that most commercial SPICE-based simulators provide for models beyond the standard ones mentioned here.

4.9 Circuit Extraction

CMOS layout drawings performed in L-Edit can be used to generate SPICE-compatible circuit file listings using the **Extract** option in the **Special** window of the Menu Bar. In essence, the extraction program translates the objects on the layout into devices, producing a text file in SPICE format. In order to use the element descriptions in a simulation, you must provide appropriate .MODEL statements.

Circuit extraction is a powerful CAD tool for

- Simulating the electrical characteristics of the actual chip layout, and,
- Providing a cross-check between the layout and the circuit schematic (known as LVS for layout versus schematic),

and is particularly useful when "fine-tuning" a design. Since the values of the parasitic capacitances and resistances depend on the layout geometries, the circuit description provided by the extraction contains critical information for high-performance design.

To use the circuit extractor, point to Special on the Menu Bar and drag the mouse down until **Extract...** is highlighted as shown. You will be prompted to provide a

The Extract operation is available in the Special window of the Menu Bar.

Special
Generate Layers...
Clear Gen'ed Layers...

DRC...
DRC Box...
Clear Error Layer...

Place and Route...

Extract...

Cross-Section...

file name for the SPICE circuit listing. L-Edit uses a default extension of **.spc** unless you specify otherwise. You are also given the option to change the extraction file if needed; for the default SCNA technology, you should ensure that the file morbn20.ext is specified. Clicking the "OK" box initiates the extraction algorithm. The actual extraction time depends on the size and complexity of the circuit and the speed (and processor) of your computer system. When L-Edit has finished writing the SPICE file, it will return you to the editing mode.

If you would like to examine the extracted file immediately, use the **Exit to DOS** option in the **File** window of the Menu Bar. This places you into a DOS shell within L-Edit. A simple way to view the extracted file is to use the command **type filename.spc**. To return to L-Edit, type **exit**.

L-Edit Example

Running Extract on the MOSFETs shown in the L-Edit examples above in Figure 4.16 produces the following SPICE file:

```
* Circuit Extracted by Tanner Research's L-Edit;
* TDB File FET VIEW 1, Cell Cell0, Extract Definition File MORBN20.EXT;

M1 12 2 13 15 PMOS L=2U W=14U
* M1 Drain Gate Source Bulk (66 18 68 32) A = 28, W = 14
.MODEL NMOS
.MODEL PMOS
.MODEL poly2NMOS
.MODEL poly2PMOS
M2 10 6 11 8 NMOS L=2U W=14U
* M2 Drain Gate Source Bulk (21 18 23 32) A = 28, W = 14
* Total Nodes: 8;
* Total Elements: 2;
* Extract Elapsed Time: 0 seconds;
.END
```

This identifies two transistors, M1 and M2. Transistor M1 is a pMOS device with connections at nodes (D,G,S,B) = (12 2 13 15), while M2 is the nMOSFET with connections at nodes (D,G,S,B) = (10 6 11 8). Note that the dimensions W=14 μm and L=2 μm (drawn values) have been correctly identified. The parameter "A" in the comment line gives the area of the gate as defined by (ACTIVE) AND (POLY).

L-Edit does not automatically identify the ground or power supply connections to the bulk or n-well regions, so the file must be edited with this information added. The .MODEL statements are included to remind you to add the information before a SPICE simulation is performed. In general, L-Edit will list all device models needed for the technology, even though you may not have every type of device in the circuit.

Parasitics

In the example above, the extracted file listing only shows MOSFETs. While this is sufficient to test the DC behavior of the network, a transient simulation requires information on the parasitic elements. L-Edit can calculate many of these contributions using information stored in the Layers windows.

Capacitance values must be entered using the Layers command. For MOSFETs created in the SCNA technology, the depletion contributions are computed using the data on the **ndiff** and **pdiff** derived layers. Interconnect contributions from POLY, METAL1, etc. may also be important, and can be entered in the same manner.

To have L-Edit calculate the parasitic capacitances, choose the "Write Node Capacitances" option in the Extract dialog box. L-Edit will then calculate linear time-invariant values based on the formula $C = C'A$, where C is the capacitance in

Farads, C' the capacitance per unit area, and A is the area. If you have not entered values for parasitics on every layer, L-Edit will give a warning for each. However, the absence of a value on a layer will not affect the calculation on any other layer. This action results in capacitor statements in standard SPICE format

> CNAME N+ N- VALUE

where CNAME will be automatically chosen, as will the node numbers N+ and N-.

It should be noted that the current version of L-Edit/Extract does not produce MOSFET statements with the drain/source size parameters AD, AS, PD, and PS, so that information on CJ, CJSW, MJ, and MJSW in a .MODEL statement will not be used when a SPICE simulation is run on a generated circuit listing. This is due to the fact that L-Edit/Extract computes parasitics directly from the data entered for each layer in the layout, not from SPICE data. The calculated depletion capacitances automatically include the areas AD and AS. However, the sidewall contributions are neglected unless the user manually adds values for PD and PS to the MOSFET element statement. This is usually only necessary in designing the circuits in critical data paths.

L-Edit/Extract can also be configured to calculate end-to-end resistances from the geometry if desired. The SCNA extract definition file morbn20.ext is set up to compute the lumped-equivalent resistance of POLY2 lines, since these are sometimes used as resistors in circuits. The resistance of other layers can be extracted by adding appropriate statements to the extract definition file, as discussed in detail in Chapter 11. Each layer requires a distinct definition statement.

4.10 References

MOSFETs are discussed in many textbooks. A few recommended references are listed below.

R4.1 Y.P. Tsividis, **Operation and Modeling of the MOS Transistor**, McGraw-Hill, New York, 1987. A definitive work on the subject.

R4.2 E. S. Yang, **Microelectronic Devices**, McGraw-Hill, 1988. This text has chapters on both the fundamental and advanced features of MOSFETs.

R4.3 J.Y. Chen, **CMOS Devices and Technology for VLSI**, Prentice-Hall, Englewood Cliffs, NJ, 1990. A good introduction to the device fabrication.

R4.4 J.P. Uyemura, **Circuit Design for CMOS VLSI**, Kluwer Academic Publishers, Norwell, MA, 1992.

4.11 Exercises

E4.1 Construct a simple n-channel MOSFET in L-Edit with $W = 20$ μm and $L'=2$ μm. Use the SCNA technology, and be sure to obey the design rules for the process. In particular, remember:

- ndiff is formed by placing ACTIVE within NESELECT. The ACTIVE edge must be at least 2λ from the NSELECT edge;
- ACTIVE CONTACTS have a size of $2\lambda \times 2\lambda$, and must be at least 2λ from an ndiff or POLY edge;
- POLY must extend beyond ndiff by 2λ;
- Metal1 is used to contact ndiff through ACTIVE CONTACT. The contact on the METAL layer must be spaced 1λ from the edge.

Follow the procedure outlined in the book for drawing the transistor. Use the cross-sectional viewer to ensure that you have used the correct layering, and run a DRC on the layout. Be sure to correct any errors that are found. Finally, perform the Extract operation and examine the SPICE listing.

E4.2 The SCNA SPICE parameters are based on an oxide thickness with a nominal value specified by TOX.

(a) Calculate the value of C_{ox} for this process. Place your answer in units of $fF/\mu m^2$.

(b) Use the information provided in the .MODEL listing to find the electron and hole mobilities μ_n and μ_p. Compare the calculated values with those in the SPICE listing (UO).

(c) An n-channel MOSFET is constructed with dimensions of L=1.5 μm and W=6 μm. Find the value of the saturation current through the device.

(d) Repeat the calculation in (c) for the case of a pMOSFET with the same dimensions.

E4.3 Construct a p-channel MOSFET with drawn dimensions of W' = 25 μm and L'=2 μm. Be sure to include the NWELL, and obey the spacing requirements.

Execute a DRC on your device, and correct any errors that are detected. Use the Cross-Sectional Viewer to examine the creation of the ndiff (n^+) regions, and verify the logic expression pdiff = (ACTIVE) AND (PSELECT) AND (NWELL) for the p-channel device.

(a) Perform the Extract operation on the device, and verify that the proper geometrical dimensions for the gate are obtained.

(b) Generate the MOSFET characteristics for I_D versus V_{DS} for V_{SG} from 0v to 5v using the extracted file and the SCNA LEVEL 2 parameters provided in the text. Ground the source and bulk electrodes for this task and use a nested .DC statement.

E4.4 Draw an n-channel MOSFET using an ACTIVE area of $20\lambda \times 10\lambda$. Then create MOSFETs with the following errors:

(a) Use a POLY gate with a width of 2λ and a total length of 12λ, but do not include any gate overhang (i.e., end one side of the POLY at the ACTIVE edge). Surround the ACTIVE polygon with an NSELECT box that has dimension of $24\lambda \times 14\lambda$. Examine the transistor using the Cross-Sectional Viewer and Extract features.

(b) Use a POLY gate with a width of 2λ, a length of 10λ, and a gate overhang of 2λ, but use an NSELECT box that has dimensions of $18\lambda \times 8\lambda$, i.e., make ACTIVE larger than NSELECT. Then study the transistor using the Cross-Sectional Viewer.

Chapter 5

CMOS Logic Circuits

CMOS is an excellent technology for designing digital integrated circuits. MOSFETs are small, allowing for high-density logic networks, and switching circuits are straightforward to design. In this chapter we will examine basic logic circuits that are used in digital VLSI chips. The discussion starts with static logic gates, which constitute a simple, but powerful, approach to designing digital logic circuits. More advanced design styles, such as dynamic circuits and BiCMOS gates, are also introduced.

5.1 Static Logic Gates

Static logic gates can be used to implement a large variety of logic operations, from basic Boolean algebra to complex logic functions. Historically, static logic gates formed the basis of CMOS networks, with complementary aspects manifest in the use of both nMOSFETs and pMOSFETs. Some of the nice characteristics of static logic gates are

- When the inputs are stable, the outputs are well-defined;
- The DC power dissipation is very small; and,
- Series and parallel MOSFET arrays can be used to construct arbitrary logic functions.

These and other properties maintain static logic circuits as a useful approach to designing digital networks in CMOS.

We will first examine the operation and design of a simple inverter circuit. This serves as a basis for constructing more complicated logic gates, and also provides a methodology for complex functions. In particular, static logic is ideally suited for implementing And-Or-Invert (AOI) and Or-And-Invert (OAI) functions directly from logical equations or from Boolean function tables.

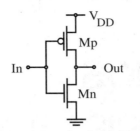

Figure 5.1.
The CMOS inverter.

5.2 The Inverter

The most basic static logic gate is the inverter circuit shown in Figure 5.1. The input is connected to the gates of both an nFET (Mn) and a pFET(Mp); the output is taken at the common drain connection. When the gates are connected in this manner, the MOSFETs constitute a **complementary pair** as discussed in the last chapter.

The operation of the inverter can be understood from either the electronic or the switch-level viewpoints. In terms of the switching behavior, it is useful to think of the power supply V_{DD} as a logic 1 rail, while the ground serves as a logic 0 rail. This results in the equivalent circuit shown in Fig. 5.2(a). Denoting the input Boolean variable by A produces an output of $\bar{A}$, as easily understood using simple reasoning. When a logic 0 is applied to the gate, the nFET switch is OPEN, while the pFET switch is CLOSED. The logic 1 (power supply) is then connected to the output. Conversely, a logic 1 input gives a CLOSED nFET switch and an OPEN pFET switch; the ground (logic 0 rail) is connected to the output. The truth table in Fig. 5.2(b) summarizes the operation of the inverter.

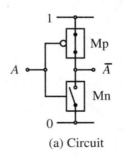

(a) Circuit

A	$\bar{A}$
0	1
1	0

(b) Truth Table

Figure 5.2.
Switch-level model for the CMOS inverter. The switch settings in (a) are for the case $A=0$.

5.2.1 Voltage Transfer Characteristics

The voltage transfer curve (VTC) of the inverter is a plot of V_{out} as a function V_{in} as shown in Figure 5.3. This is derived by simply equating the drain currents of the transistors while noting the operational mode of each (cutoff, saturation, or non-saturation). The VTC defines the DC behavior of the circuit, and neglects all transient switching effects. The transfer curve illustrates the following important points about the operation of the inverter:

- The largest output voltage is V_{DD}, which is an ideal logic 1 voltage;
- An output voltage of V_{DD} can be attained by a range of input voltages close to $0v$.;
- The smallest output voltage is $0v$, which is an ideal logic 0 voltage;
- An output voltage of $0v$ can be attained by a range of input voltages close to V_{DD}.

The logic 0 and logic 1 input voltage ranges are shown explicitly on the curve.

The VTC also shows the unity-gain line defined by $V_{out} = V_{in}$. The intersection of the transfer curve with the unity gain line is denoted by V_I, which is used to denote the inverter switching voltage. The interpretation of V_I is straightforward: it represents the border between a logic 0 and a logic 1 input voltage range. V_I is thus called the inverter threshold voltage.

5.2.2 Transient Properties

The switching characteristics of the inverter are summarized in Figure 5.4 for the case of a square wave input voltage $V_{in}(t)$. Physically, logic delays originate from the finite time required to charge and discharge the output capacitance C_{out} of the inverter. Figure 5.5 shows the transient circuit with C_{out} for the idealized case of a step input voltage. The switching time intervals are computed by analyzing the output circuit for high-to-low and low-to-high transitions. The important switching

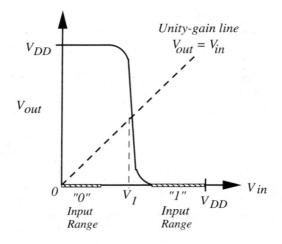

Figure 5.3.
The inverter voltage transfer curve (VTC)

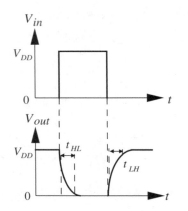

Figure 5.4
Definition of inverter switching times.

intervals are given by

- t_{HL}: The output **high-to-low time** [1]. This is the time required for the output voltage to change from $0.9V_{DD}$ to $0.1V_{DD}$ [2].

- t_{LH}: The output **low-to-high time** [3]. This is the time required for the output voltage to change from $0.9V_{DD}$ to $0.1V_{DD}$.

These switching times can be used as the basic measure for the speed of the inverter. For example, consider the sum $(t_{HL} + t_{LH})$. This is the minimum time interval needed for the inverter to make a full transition from a logic 1 to a logic 0, and back to a logic 1. This is used to define the maximum switching frequency f_{max} by means of

$$f_{max} = \frac{1}{t_{HL} + t_{LH}} ,$$

(5. 1)

and represents the upper theoretical limit of the circuit response.

The propagation delay time t_p is a measure of the average delay through the gate. It can be estimated by

$$t_P = \left(\frac{1}{2}\right)(t_{PLH} + t_{PHL}) ,$$

(5. 2)

where t_{PHL} and t_{PLH} are the propagation delays for a high-to-low, and a low-to-high transition, respectively, shown in Figure 5.6. These times are defined between 50% points as shown. Propagation delays are often used in logic simulations to estimate

---

[1] t_{HL} is also known as the fall or discharge time.

[2] The voltages $.1V_{DD}$ and $.9V_{DD}$ represent the 10% and 90% values of the output logic swing, respectively.

[3] Other names for t_{LH} include the rise time and the charge time.

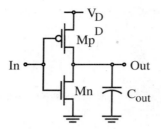

Figure 5.5.
Inverter output capacitance C_{out} as a load.

the average delay time through the gate.

5.2.3 Design

Inverter design is based on both the DC and the transient characteristics. The design variables for the circuit are the MOSFET aspect ratios $(W/L)_n$ and $(W/L)_p$. Device sizes are chosen to achieve the desired properties.

Let us first examine the DC transfer curve. The transition voltages are set by the transconductance ratio

$$\beta_R = \frac{\beta_n}{\beta_p} = \frac{k'_n \left(\dfrac{W}{L}\right)_n}{k'_p \left(\dfrac{W}{L}\right)_p}. \tag{5.3}$$

The simplest approach to illustrating the design is to examine the inverter threshold voltage V_I, which occurs when the input and output voltages are equal:

$$V_{in} = V_{out} = V_I. \tag{5.4}$$

Graphically, this corresponds to the point where the VTC intersects the unity gain line $V_{out}=V_{in}$, as discussed earlier. A quick examination of the devices shows that

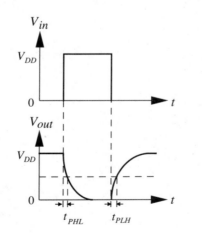

Figure 5.6.
Propagation delay times.

both MOSFETs are saturated at this point. Equating currents and solving gives

$$V_I = \frac{\sqrt{\beta_R} V_{Tn} + (V_{DD} - |V_{Tp}|)}{1 + \sqrt{\beta_R}} \qquad (5.5)$$

for the inverter threshold voltage. This shows that the relative value of β_n to β_p sets the DC transfer characteristics of the inverter.

The inverter threshold voltage V_I can be adjusted using the ratio (β_n/β_p). For the case $(\beta_n/\beta_p) = 1$ and $V_{Tn} \approx |V_{Tp}|$, the equation shows that $V_I \approx (V_{DD}/2)$. This places the transition at the midpoint of the VTC. If $(\beta_n/\beta_p) > 1$, then V_I is reduced, while $(\beta_n/\beta_p) < 1$ increases its value. These are shown in Figure 5.7.

The transient switching times depend on the values of the device resistances as combined with the output capacitance C_{out}. Figure 5.8 shows the circuits for the charging and discharging events. For a charging event, the pFET is on and conducts current from the power supply to charge the output capacitor. The time constant for the low-to-high time can be written as

$$\tau_p = R_p C_{out} = \frac{C_{out}}{\beta_p (V_{DD} - |V_{Tp}|)}, \qquad (5.6)$$

where R_p is the equivalent pMOSFET resistance. The actual value of t_{LH} is a little larger than τ_p, so $t_{HL} \approx 2\tau_p$ is used as a reasonable first-order estimate. Similarly, discharging the output capacitor through the nFET is characterized by the time constant

$$\tau_n = R_n C_{out} = \frac{C_{out}}{\beta_n (V_{DD} - V_{Tn})}, \qquad (5.7)$$

with R_n as the equivalent nMOSFET resistance. The transition time t_{HL} can be estimated to first order using $t_{LH} \approx 2\tau_n$. The propagation delay can be estimated to first order from

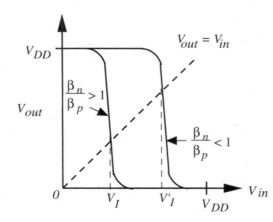

Figure 5.7.
Effect of varying the β-ratio on the VTC.

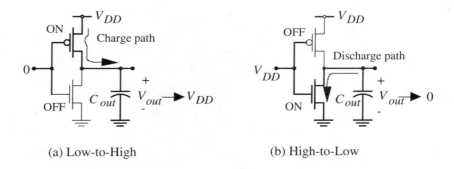

(a) Low-to-High   (b) High-to-Low

Figure 5.8. Charging and discharging of the output capacitance.

$$t_p \approx \frac{(\tau_n + \tau_p)}{2}, \qquad (5.8)$$

which approximates the propagation charge and discharge times as being single time constants. It is important to remember that the values predicted by these equations are only first-order estimates.

The design equations presented above illustrate that the DC characteristics can be adjusted by varying the ratio of β_n to β_p, while the transient times are largely dependent on the actual values of β_n and β_p. The transient analysis is complicated by the fact that C_{out} contains terms due to parasitic MOSFET capacitances that are dependent on the values chosen for the aspect ratios. A simple estimate is to write

$$C_{out} = C_{GDn} + C_{GDp} + C_{DBn} + C_{DBp} + C_{Line} + C_{in} \qquad (5.9)$$

where C_{in} is the input capacitance into the next stage. Note in particular that the capacitances C_{GD} and C_{DB} increase with the channel length W. Thus, increasing the aspect ratio *(W/L)* of a MOSFET decreases the resistance, but increases its internal capacitance at approximately the same rate. If C_{out} is dominated by external load capacitance contributions ($C_{Line} + C_{in}$), then increasing the device size will in fact decrease the switching times. Otherwise, the switching time is limited by the process technology factors.

5.2.4 Layout

Circuit layout and physical design revolve around two basic ideas. First, we want to design the circuit so that it has the proper electrical characteristics. Second, we need to place the transistors and the interconnect lines in a manner that results in a reasonably compact structure and also allows us to route other connections as needed. Many variations are possible.

A basic inverter layout is shown in Figure 5.9. This places the transistors and wiring in a one-to-one correspondence with the circuit drawn in Figure 5.1. Note that the transistor dimensions W' and L' are the drawn values. In this particular design, the pFET has been made larger than the nFET: $(W/L)_p > (W/L)_n$. However,

---

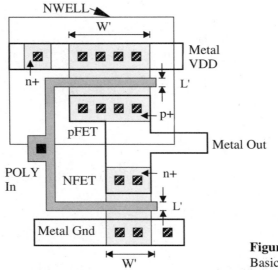

Figure 5.9.
Basic inverter layout.

the size ratio is less than 2, so $\beta_n > \beta_p$ for the design.

A different approach for layout is illustrated in Figure 5.10. In this example, the MOSFETs are constructed using horizontal ACTIVE regions. Both the POLY input and all METAL connections are vertical, with contacts placed to provide the proper wiring arrangement. This inverter has been designed with equal aspect ratios, so that $\beta_n > \beta_p$, with the nFET device transconductance larger than the pFET value by a factor of (k'_n/k'_p).

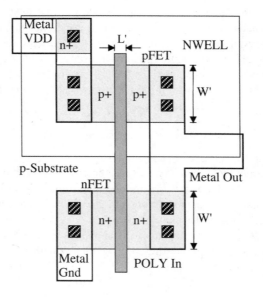

Figure 5.10.
Alternate inverter layout.

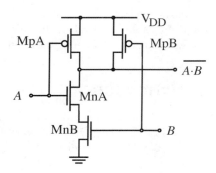

Figure 5.11.
CMOS NAND2 logic gate.

5.3 NOR and NAND Gates

The inverter circuit can be easily modified to implement the NAND and NOR functions. A NAND2 gate[1] is shown in Figure 5.11; it consists of two nFETs in series and two pFETs in parallel. The nFETs provide a path from the output to ground if and only if both $A=1$ **AND** $B=1$; otherwise, at least one of the nFETs is OFF and the output has a strong conduction path to the power supply V_{DD} through one of the pFETs. The formation of the NAND operation is easily understood. When $AB=1$, the output is a logic 0 output; note that the circuit structure automatically implements the NOT-AND operation. If either input is 0, then a conduction path between the power supply and the output exists, bringing the output to a logic 1 value.

A NOR2 gate has a similar structure as can be seen in Figure 5.12. In this case, the two nFETs are placed in parallel, while the two pFETs are in series. If either $A=1$ **OR** $B=1$ (or both), then at least one of the nFETs is conducting and the output voltage is $0v$. The only time the output voltage is high (V_{DD}) is when both $A=0$ and $B=0$, since this is the only case where both pFETs are conducting. Denoting the OR operation by "+", we see that when $(A+B)=1$, the output is 0. Thus, negation is automatic from the circuit structure of the nMOSFETs, and we obtain the NOT-OR function.

The NAND and NOR gate examples give the basis for implementing more com-

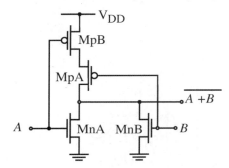

Figure 5.12.
CMOS NOR2 logic gate.

---

[1] The notation NAND2 implies a 2-input NAND gate.

---

plex logic functions. Note that the nFETs and pFETs have complementary arrangements: when the nFETs are in parallel, the pFETs are in series, and vice-versa. These simple observations are sufficient to extend the basic circuit to more general gate design.

5.4 Logic Formation

CMOS allows us to construct logic gates that implement functions more complex than the basic NOT, NAND, and NOR operations. This is one reason why CMOS is successful for VLSI circuits. Generalized static logic gate design is based on the use of complementary pairs of MOSFETs, giving the general circuit structure shown in Figure 5.13. Important features of the topology are as follows.

- Every input variable is connected to both an nFET and a pFET;

- Two logic arrays are used to implement the logic function. One array consists of nFETs with the logic block connecting the output to ground. The other array is made up of pFETs, and constitutes a logic block connected from the output to V_{DD};

- When the inputs are stable, only one logic block (nFET or pFET) is closed (i.e., conducts from the top to the bottom).

We note that an N-input gate requires 2N transistors in this scheme (one nFET and one pFET for each input). The use of complementary logic blocks gives general static logic gates many of the desirable properties of the inverter, such as low DC power dissipation.

Static logic gates can implement AOI (And-Or-Invert) and OAI (Or-And-Inverter) logic functions in a straightforward manner. These correspond to inverted canonical SOP (Sum of Products) and POS (Product of Sums) forms, and are therefore very useful for implementing logic designs. The basic rules of logic formation can be deduced by studying the NAND and NOR gate topologies. We can summarize the rules by the following statements:

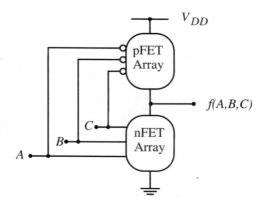

Figure 5.13.
General CMOS static logic gate.

- Series-connected nMOSFETs give the NAND operation;
- Parallel-connected nMOSFETs give the NOR operation.

These rules also apply to groups of nFETs that implement individual functions. For the p-channel transistors, the logic formation is given by

- Series-connected pMOSFETs to implement the NOR operation;
- Parallel-connected pMOSFETs to implement the NAND operation.

To apply these rules, design the nFET array first. Each transistor is assigned an input variable, and is arranged according to the rules above. For example, two series-connected nFETs with inputs X and Y implement the function

$$NOT(X \text{ AND } Y) = NAND(X,Y)$$

when placed in the circuit. If the two nFETs are in parallel, then we have

$$NOT(X \text{ OR } Y) = NOR(X,Y)$$

for the pair. The structure of the pFET logic array is just the dual of the nFET connections. In other words, nFETs in series require that the corresponding pFETs be in parallel, and vice-versa.

As an example, consider the function

$$F = \overline{AB + C}. \tag{5.10}$$

To have an output of $F=0$ requires that $(A \text{ AND } B)=1$ OR $(C=1)$. For the nFET array, this means that we need three transistors with inputs A, B, and C, such that the A - and B -FETs are in series, and the C- FET is in parallel with this group. Figure 5.14 shows the logic gate structure obtained by directly applying these rules.

As another example, let's look at the exclusive-OR (XOR) function

$$F = A \oplus B. \tag{5.11}$$

This can be expressed in AND-OR form as

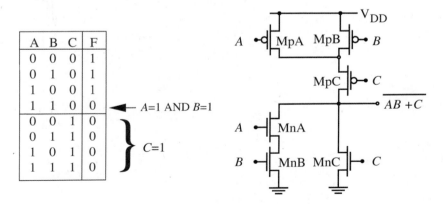

A	B	C	F	
0	0	0	1	
0	1	0	1	
1	0	0	1	
1	1	0	0	← $A=1$ AND $B=1$
0	0	1	0	
0	1	1	0	
1	0	1	0	$C=1$
1	1	1	0	

Figure 5.14. Example of a CMOS AOI logic gate.

A	B	$A \oplus B$
0	0	0
0	1	1
1	0	1
1	1	0

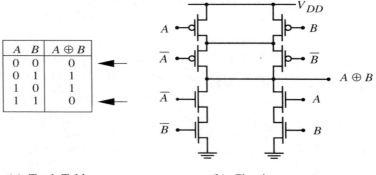

(a) Truth Table (b) Circuit

Figure 5.15. Exclusive-OR gate.

$$F = A \cdot \bar{B} + \bar{A} \cdot B. \tag{5.12}$$

To create the AOI invert form needed for the logic gate, we simply note that

$$F = \overline{(A \cdot B + \bar{A} \cdot \bar{B})}, \tag{5.13}$$

which leads directly to the logic gate shown in Figure 5.15. Since the exclusive-NOR (XNOR or Equivalence) function can be expressed as

$$F = A \cdot B + \bar{A} \cdot \bar{B}, \tag{5.14}$$

the XNOR gate has the same structure but with B and $\bar{B}$ interchanged. The resulting circuit is shown in Figure 5.16.

A	B	$\overline{A \oplus B}$
0	0	1
0	1	0
1	0	0
1	1	1

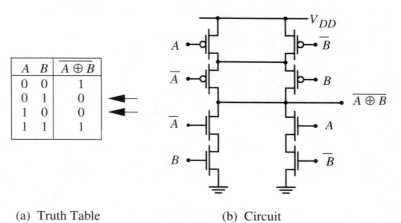

(a) Truth Table (b) Circuit

Figure 5.16. Exclusive-NOR (equivalence) gate.

CMOS Logic Circuits

5.5 Layout of Complex Logic Gates

It is seen from these examples that static logic gates for complex logic functions can be constructed by straightforward application of the logic formation rules. Since the rules presented here are based on series and parallel combinations of MOSFETs, the layout techniques can be divided into three main groups:

- Series-connected MOSFETs;
- Parallel-connected MOSFETs; and
- Input and output wiring.

In addition, the wiring to ground and VDD is always required.

Series-connected MOSFETs are relatively simple to design since n^+ and p^+ regions can be shared between two transistors. Figure 5.17 shows three series-connected nFETs. Since the drain and source electrodes are not defined until the voltages are applied, a common n^+ region serves as either a drain or source as required. The series chain can be constructed by simply placing three poly gates in parallel as shown. The transient delay through the chain of transistors is important to estimate. Since the transistor resistances are in series, the RC time constant of the group can be a limiting factor in the switching time.

Wiring MOSFETs in parallel is also straightforward. The simplest approach is to achieve the parallel connections using n^+ or p^+ regions in conjunction with metal interconnect routing. Figure 5.18 shows a layout technique for three pFETs in parallel between V_{DD} and the output.

Series-parallel logic can be created with many different layout schemes. Figure 5.19 shows both a NAND2 and a NOR2 gate. This is an interesting example of the symmetry between dual logic gates[1]. In a NAND2 circuit, two nFETs are in series and two pFETs are in parallel. A NOR2 gate is exactly opposite, with two nFETs in parallel and two pFETs in series. Comparing the layouts shows how this dual nature arises in the layout. Consider the METAL pattern on the NAND2 gate. If you perform a vertical "flip" on the layer pattern, then you obtain the NOR2 gate! Symmetries of this type are common in series-parallel logic circuits.

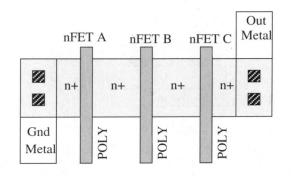

Figure 5.17.
Layout of series-connected MOSFETs.

---

[1] Recall that dual means OR replaced by AND, and vice-versa. Hence, the NAND2 is the dual of the NOR2.

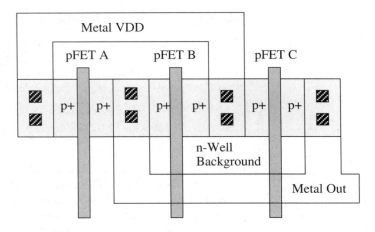

Figure 5.18. Layout of parallel-connected MOSFETs.

5.5.1 Layout Hints

Since the number of MOSFETs in a CMOS logic circuit can get large, it is useful to develop a basic methodology to reduce the work needed to create the layout. The comments below are meant only as a recommended approach for those new to the field. After a little practice, you will find that CMOS layout becomes almost second nature, and for a few[1], even a little fun!

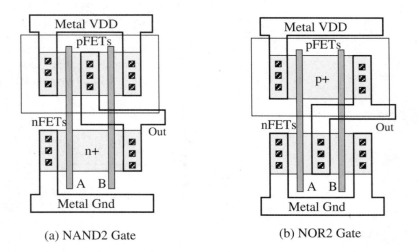

(a) NAND2 Gate   (b) NOR2 Gate

Figure 5.19. Example layouts for NAND2 and NOR2 gates.

---

[1] Those of us known as "nerds."

The starting point is the schematic diagram. This tells us (i) the placement of the transistors, and (ii) the interconnect scheme. All static logic gates have power supply (V_{DD}) and ground (or V_{SS}) connections, and these are useful reference polygons for planning the layout. Both are created using parallel Metal1 (or higher Metal layers) lines according to the chip power distribution scheme[1].

Arranging MOSFETs and providing the proper wiring can be the most challenging task in building a logic circuit. It is often useful to introduce the concept of **stick diagrams** to aid in this task. A stick diagram is an alternate representation of the circuit, somewhere in between a layout drawing and the electronic schematic. To create a stick diagram, we introduce various lines (the "sticks") to represent each layer; the width of the line is not important, only the connection. The lines can be color coded according to the layout editor screen (e.g., a red line to represent an L-Edit POLY layer), or, the layers can be distinguished using lines with different characteristics: solid, dashed, bold, etc. Since we have opted to minimize the cost of this book/software package, color printing was deemed too expensive. We will therefore use the latter approach where different types of lines are used. However, since a set of color pencils or pens is quite reasonable, you should use colors to code the lines in your own work, as it is much easier to "see" the interconnects.

Consider the basic stick definitions in Figure 5.20(a). Every layer is represented by a particular type of line, so that we may construct a stick diagram for any circuit. To create a MOSFET, we simply cross n^+ with POLY as shown in Figure 5.20(b). Contacts and VIAS are denoted using an "×".

Stick diagrams are used to help plan the layout of a circuit. Figure 5.21 shows how stick diagrams may be used to construct a simple inverter. Two layout approaches are shown. MOSFETs, whether n-channel or p-channel, are represented by POLY crossing N+ or P+ to simplify the drawings. Metal1 is used for the power supply and ground connections. In both circuits, the input (*In*) is applied to the POLY layer, while the output (*Out*) is obtained from a Metal1 layer. It is seen that the stick diagrams provide the basic topological features needed for the layout.

As another example, consider the function

POLY:	
N+/P+:	
METAL1:	
CONTACT: ×	

    (a) Definitions   (b) MOSFET

Figure 5.20. Basic stick diagram definitions.

[1] Power distribution is discussed in Chapter 6.

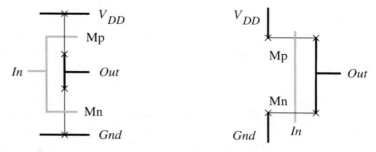

Figure 5.21. Stick diagrams for the CMOS inverter.

$$F = \overline{A\,(B + C)}\ . \tag{5.15}$$

Figure 5.22 shows two possible stick diagrams for the CMOS logic gate. The pMOSFETs are the transistors (POLY over P+) that are in the upper half of the circuit (near V_{DD}), while the nMOSFETs are clustered around the ground. The interconnect problem in VLSI starts to become apparent in this circuit. In order to connect the gates, Metal1 must be used to cross over POLY or N+/P+ regions. Stick diagrams are particularly useful for planning interconnect strategies.

Translating a stick diagram to a physical layout is straightforward. Each line represents a particular material layer that must be replaced by a polygon of appropriate size and shape. The simplest approach is to initially represent MOSFETs using only POLY over ACTIVE regions. Place the transistors according to the topology specified by the stick diagram; the size of an ACTIVE area is chosen according to the electrical (W/L) design. Group nMOSFETs and pMOSFETs together as they are added to the layout, then surround the transistors with NSELECT or PSELECT-NWELL polygons to create nMOSFETs or pMOSFETs, respectively. Don't worry about changing the layout as the drawing evolves. Stick diagrams are only guides, and do not account for the shapes and sizes of the polygons needed in the physical design.

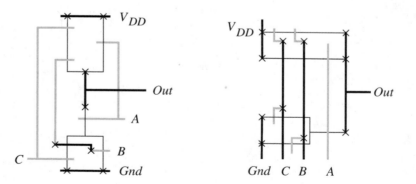

Figure 5.22. Stick diagram for an AOI logic gate.

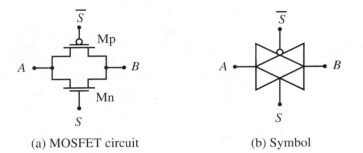

(a) MOSFET circuit   (b) Symbol

Figure 5.23. CMOS transmission gate (TG).

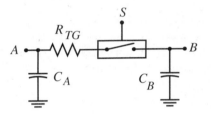

Figure 5.24.
Switching model of a
CMOS transmission gate.

5.6 CMOS Transmission Gates

A CMOS transmission gate (TG) consists of an nFET in parallel with a pFET as shown in Figure 5.23(a). The simplified logic symbol in Figure 5.23(b) will be used to represent the combination in logic diagrams. The transistors are controlled by complementary signals S and $\bar{S}$. When $S=0$, both FETs are in cutoff, and the TG may be modelled as an open switch. Setting the control signal to $S=1$ turns on both MOSFETs, allowing conduction. In this case, the TG acts like a closed switch between the left and right sides. Logically, the function of the TG can be written as

$$S=1: B \leftarrow A, \tag{5.16}$$

i.e., when $S=1$, then A is transferred through the TG to the output B. When $S=0$, the switch is open, and the relationship between the input and the output is not defined. This assumes that logic propagates from the left to the right. However, the CMOS transmission gate is bi-directional, and can transmit either direction.

It is important to note the existence of parasitic resistance and capacitance in the MOSFETs when creating an electrical model of the transmission gate. Figure 5.24 shows the equivalent circuit with the parasitics included. When the TG is used in a circuit, it acts as a passive switch that intrinsically slows the response due to the parasitic RC structure.

5.6.1 Layout

Layout of a CMOS TG is complicated by the fact that the nFET and pFET transistors reside in opposite-polarity background regions. Assuming a p-substrate, this means that we must include an n-well region for the pFET. Since the design rule spacings between doped regions in the silicon are relatively large, the integration density may be reduced.

An example of a TG layout is shown in Figure 5.25. This is simply the inverter layout from Figure 5.10 with modified metal connections. The electrical characteristics are established by two primary features of the layout. First, the aspect ratio of the transistors gives the resistances R_n and R_p for the individual FETs. The TG resistance is then estimated from

$$R_{TG} = (R_n \| R_p) = \frac{R_n R_p}{R_n + R_p}.$$ (5. 17)

Caution must be exercised when using this value since R_{TG} is a linear approximation for a pair of nonlinear devices. The capacitance contributions of the gates are determined by the dimensions of the transistors. A simple estimate for the capacitors in the simplified switch model is

$$C_A = C_B \approx (C_{GDn} + C_{DBn}) + (C_{GSp} + C_{SBp}) .$$ (5. 18)

This takes into account both devices. (The labelling of the drain and source is opposite for the nFET and pFET.)

5.6.2 TG Logic Circuits

Since transmission gates act as voltage-controlled switches, they can be used to implement Boolean switching functions. The simplest implementations are those

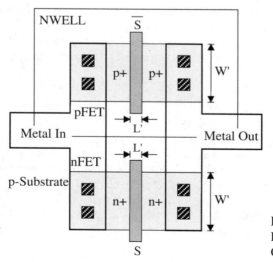

Figure 5.25.
Example layout for a CMOS transmission gate.

that can be expressed in sum-of-product (SOP) form. These circuits are based on the simple multiplexer shown in Figure 5.26. Two data lines D_0 and D_1 form the inputs. The output T is determined by the switching control bit S. The operation of the circuit is straightforward. If $S=0$, then TG0 acts as a closed switch while TG1 is an open switch. If $S=1$, then TG0 is open and TG1 is closed. This action is described by the Boolean expression

$$T = D_0 S + D_1 \bar{S}. \qquad (5.19)$$

The circuit is designed to avoid the situation where the output is floating, since that would give an undefined output condition.

Many logic functions can be placed into canonical SOP form and implemented using TG-based switched logic networks. For example, recall the XOR and XNOR functions

$$F(A, B) = A\bar{B} + \bar{A}B$$
$$\bar{F}(A, B) = AB + \bar{A}\bar{B} \qquad (5.20)$$

The TG logic circuits are shown in Figure 5.27. Both are obtained directly from the 2-input MUX network. Half-adder and full-adder circuits can be constructed using these circuits as building blocks.

Transmission gates can be used to construct a simple register as shown in Figure 5.28. In this circuit, the load control bit LD determines the operation. When $LD=1$, the input transmission gate $TG0$ is closed while $TG1$ is open. Data bit D then enters the circuit. Switching the load bit to a value of $LD=0$ opens $TG0$ and closes $TG1$. The inverters then form a closed-feedback loop, holding the data bit D until LD is switched back to a value $LD=1$.

Cascading two register circuits yields the clock-controlled D-type flip-flop shown in Figure 5.29. Data transfer is controlled by the clock signal ϕ. A condition of $\phi=0$ allows data into the first register; when the clock switches to $\phi=1$, the data bit is transferred to the second register, and is available at the output. Logically, the

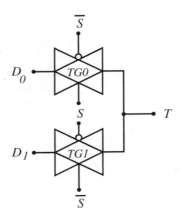

Figure 5.26.
Transmission gate
2:1 multiplexer circuit.

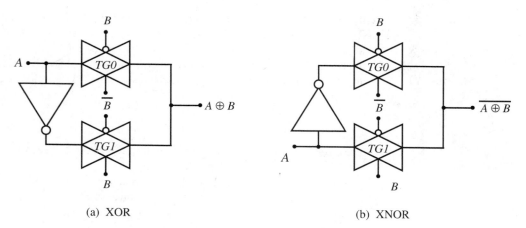

(a) XOR   (b) XNOR

Figure 5.27. TG-based XOR and XNOR functions.

operation can be written as

$$Q(t+1) = D(t), \tag{5.21}$$

which is the defining relationship for a D-type flip-flop. Replacing selected inverters with 2-input NAND or NOR gates gives additional inputs that can be used to control the contents. For example, a clear CLR input can be added to initialize the contents to logic 0 values.

Clock-controlled TGs can also be used to synchronize data flow in static logic circuits. A basic shift register is shown in Figure 5.30. Bit transfers from one stage to the next occur when the TG is switched ON. By alternating the clocking signals ϕ and $\bar{\phi}$ as shown, data are shifted on half clock cycle intervals. The technique of using clocked transmission gates to time data transfer can easily be extended to arbitrary networks.

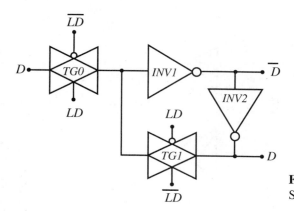

Figure 5.28.
Storage register.

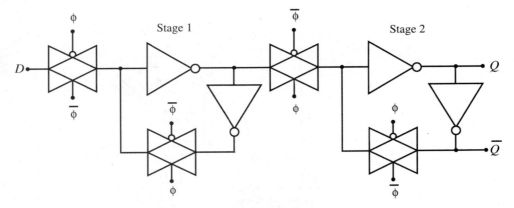

Figure 5.29. D-type flip-flop.

5.7 Dynamic Logic Circuits

Dynamic CMOS logic circuits use FET switches and parasitic capacitors to implement logic functions. Clock signals are used to synchronize the flow of charge in the network. In a dynamic circuit, the output is only valid for a short period of time, even if the inputs are stable.

A basic dynamic NAND2 gate is shown in Figure 5.31. Clocking is controlled by transistors Mp and Mn, while nFETs M1, M2, and M3 perform the logic. Two distinct operational phases can be defined for this circuit. Assume that, initially, the clock has a value $\phi=0$, so that Mp is ON and Mn is OFF. This constitutes the **pre-charge** interval. During this time, the output capacitor C_{out} is charged to a voltage V_{DD} through Mp.

A clock state of $\phi=1$ defines the **evaluate** phase. During evaluation, Mp is OFF and Mn is ON, and the behavior of the circuit depends on the input variables. If at least one of the inputs A, B, or C is a logic 0 value, then the nFET logic chain made up of M1, M2, and M3 acts like an open circuit. The output voltage remains at V_{DD}, corresponding to a logic 1 level. If all of the inputs are 1, then the logic chain acts

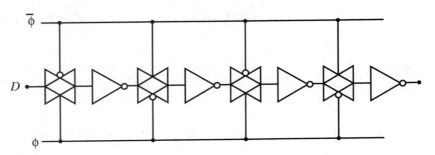

Figure 5.30. Shift register.

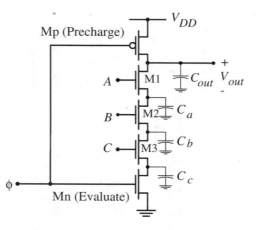

Figure 5.31.
Dynamic 3-input
NAND gate.

like a closed circuit, and C_{out} discharges to a voltage $V_{out}=0$ corresponding to a logic 0 value. The output is said to undergo a **conditional discharge**. It is seen that this implements the NAND function

$$F = \overline{A \cdot B \cdot C}. \tag{5.22}$$

Note, however, that the output is only valid during the evaluate interval when the clock has a value of $\phi=1$. The dynamic nature of this type of logic circuitry is apparent.

5.7.1 Charge Sharing and Charge Leakage

Two problems arise in the operation of the basic dynamic logic stage. The first is that of **charge sharing**, a mechanism that reduces the value of the output voltage below its precharge level. To understand this property, suppose that the output is charged to a voltage of V_{DD} when the $A=0$ exists at the input to transistor M1. The total charge in the circuit is then

$$Q_T = C_{out}V_{DD}, \tag{5.23}$$

and is stored entirely on the output capacitor. Now suppose that the inputs are switched to $(A,B,C) =(1,1,0)$; MOSFETs M1 and M2 are on, while M3 remains off. Since a conduction path exists between the output node (C_{out}) and capacitors C_a and C_b, current flow is initiated and the charge is shared among the three capacitors. Eventually, equilibrium is established. Since the capacitors are all in parallel, they all have the same final voltage V_f, so that the charge is now distributed according to

$$Q_T = C_{out}V_f + C_aV_f + C_bV_f. \tag{5.24}$$

Since the total charge Q_T is a constant, equating the two expressions gives

$$V_f = \frac{C_{out}}{C_{out} + C_a + C_b} V_{DD}, \tag{5.25}$$

clearly showing that $V_f < V_{DD}$. To insure that the output voltage remains large enough to still be interpreted as a logic 1 requires that the capacitor values satisfy the relation

$$C_{out} \gg C_a + C_b. \tag{5.26}$$

Since the capacitances are determined by the layout geometries, care must be taken to ensure that charge sharing does not reduce the voltage.

The second problem with this circuit is that the output voltage only remains valid for a short amount of time due to **charge leakage** from the output node. This is primarily due to reverse leakage currents across the pn junctions at the drain-bulk and source-bulk regions. Consider a MOSFET that is holding charge on a capacitor C_L as shown in Figure 5.32(a). Ideally, the zero-gate voltage would give $I_D=0$. However, a reverse-biased pn junction admits a leakage current I_R that cannot be eliminated. The charge leakage equation is equivalent to the basic capacitor relation

$$I_R = -C_L \frac{dv_L}{dt}. \tag{5.27}$$

A simple approximation is to assume that both I_R and C_L are constant. Integrating with the initial voltage $V_L(0)=V_0$ gives the voltage as

$$V_L(t) \approx V_0 - \frac{I_R}{C_L} t. \tag{5.28}$$

Figure 5.32(b) shows a more realistic decay of the voltage, with the linear approximation included for reference. The time t_1 that the MOSFET can hold the voltage above a value $V_1=V_L(t_1)$ is given by

$$t_1 \approx \frac{C_L}{I_R}(V_0 - V_1). \tag{5.29}$$

For reverse currents on the order of *picoamperes* $(10^{-12}\ A)$ and capacitance values on the order of *femtofarads* $(10^{-15}\ F)$, this implies that a high (logic 1) voltage can be held for a time on the order of a *millisecond* $(1\ ms = 10^{-3}\ s)$. In other words, the output voltage is only valid for a short period of time. This is the origin of the adjective "dynamic" for this type of circuit.

5.7.2 Layout Considerations

Dynamic logic circuits require less interconnect than static logic circuits since the inputs are only connected to nFETs. However, the problems of charge sharing and charge leakage make the operation of the circuit much more sensitive to the layout geometries. This arises because of parasitic depletion capacitance between two series-connected MOSFETs is proportional to the area A of the n^+ regions.

A sample layout for a dynamic NAND3 circuit is shown in Figure 5.33. This lay-

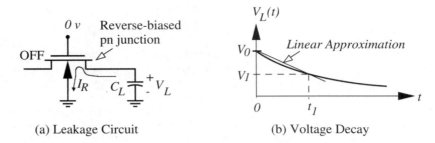

(a) Leakage Circuit   (b) Voltage Decay

Figure 5.32. Charge leakage through a MOSFET.

out style is based on the inverter example of Figure 5.10. The clocked-controlled transistors Mn and Mp share a common gate. The logic MOSFETs M1, M2, and M3 are series-connected using a single n$^+$ region with individual POLY gates. The parasitic capacitance between two logic FETs is estimated as having depletion and gate-channel contributions.Using the geometry in the drawing, we can write

$$C_a \approx C_{j0}XW' + C_{jsw}(2X + 2W') + C_{ox}L'W' \tag{5.30}$$

and $C_a \approx C_b \approx C_c$. The output capacitance consists of similar contributions from Mp and MA, and also the line and input capacitance of the next stage. The trade-offs in the geometry arise from the observation that increasing W allows more current flow, but increases the internal capacitances of the circuit. Both charge sharing and charge leakage depend on the areas. In practice, the easiest design route is to use

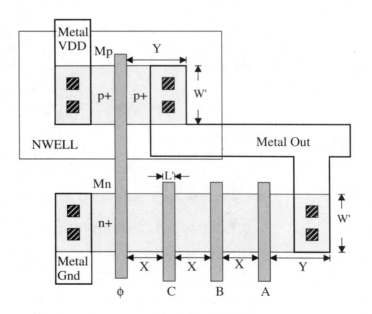

Figure 5.33. Dynamic circuit layout example.

reasonable dimensions, simulate the circuit, and then adjust the sizes as necessary to meet the design specifications.

5.7.3 Other Dynamic Circuit Styles

CMOS admits to several dynamic circuit design styles. Included in the list are Domino Logic (with several variations), NORA (no-race) logic, and Zipper circuits. All of these examples are clock-controlled, and use the Precharge and Evaluate cycles discussed above. More details can be found in the references at the end of the chapter.

5.8 BiCMOS Circuits

BiCMOS circuits employ both MOS and bipolar transistors with the philosophy that

- CMOS provides high-density logic, while
- Bipolar transistors improve the switching speeds.

Several commercial chip designs are based on a BiCMOS technology.

5.8.1 Bipolar Junction Transistors

Bipolar junction transistors (BJTs) have three terminals that are named the emitter, the base, and the collector. The circuit symbol for an npn BJT is shown in Figure 5.34[1]. Current flow through a bipolar transistor is quite different from that found in a MOSFET. In general, the base-emitter voltage V_{BE} and the base-collector voltage V_{BC} control the operation of the device. There are four regions of operation for the bipolar transistor depending on the junction bias; these are summarized in Table 5.1.

TABLE 5.1 Operating regions of a bipolar transistor.

Region of Operation	Voltages	Characteristics
Forward Active	$V_{BE}>0, V_{BC}<0$	High gain, amplification
Saturation	$V_{BE}>0, V_{BC}>0$	I set by external circuit
Cutoff	$V_{BE}<0, V_{BC}<0$	Only leakage currents
Reverse Active	$V_{BE}<0, V_{BC}>0$	Low gain

The DC characteristics can be described using the Ebers-Moll equations for the emitter current I_E and the collector current I_C in the form

---

[1] Since pnp transistors are not very common in BiCMOS designs, we will restrict our attention to analyzing the npn device.

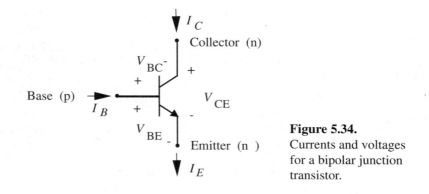

Figure 5.34.
Currents and voltages for a bipolar junction transistor.

$$I_E = I_{EX}[exp(V_{BE}/\phi_{th}) - 1] - \alpha_R I_{CS}[exp(V_{BC}/\phi_{th}) - 1]$$

$$I_C = \alpha_F I_{EX}[exp(V_{BE}/\phi_{th}) - 1] - I_{CS}[exp(V_{BC}/\phi_{th}) - 1] \quad , \tag{5.31}$$

where I_{ES} and I_{CS} are the emitter and collector saturation currents, respectively, α_F (α_R) is the forward (reverse) current gain, and $\phi_{th}=(kT/q)$ is the thermal voltage. The base current is given by $I_B=I_E-I_C$.

Forward active operation is obtained with $V_{BE} >0$ and $V_{BC} <0$, i.e., the base-emitter junction is forward-biased, and the base-collector junction is reverse-biased. In this case, the equations reduce to the form

$$I_E \approx I_S exp(V_{BE}/\phi_{th})$$

$$I_C \approx \alpha_F I_S exp(V_{BE}/\phi_{th}) \quad , \tag{5.32}$$

where I_S is the saturation current, and reverse leakage currents have been neglected. These illustrate the important point that bipolar transistors are characterized by an exponential current-voltage relation for $I_C(V_{BE})$. MOSFETs, on the other hand, have an (approximate) square-law relation behavior for $I_D(V_{GS})$. This shows that bipolar transistors are more sensitive to the input voltage, which in turn makes them faster switches. In addition, bipolar transistors can handle much higher current levels, enabling them to charge or discharge capacitors faster. BiCMOS circuits are constructed to take advantage of these characteristics.

5.8.2 Layout

The surface view of an integrated bipolar transistor is shown in Figure 5.35(a). Although the terminals are taken from the top of the structure, the npn layering is downward as shown by the cross-sectional view in Figure 5.35(b), making this a vertical device. The saturation current I_S is proportional to the area A_E of the emitter. Since the transistor action takes place under the emitter n^+ region, A_E can be approximated as the emitter surface area.

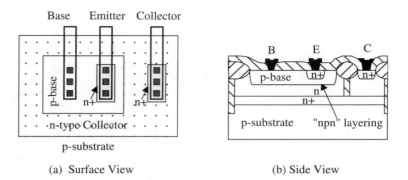

Base   Emitter   Collector

p-base

n+   n+

·n-type Collector· · · ·

p-substrate

(a) Surface View

B   E   C

p-base   n+   n+

n

n+

p-substrate   "npn" layering

(b) Side View

Figure 5.35. Geometry of an integrated bipolar transistor.

A BiCMOS process is designed to provide nFETs, pFETs, and bipolar transistors in a single substrate; some also provide pnp transistors. The default SCNA technology that is loaded when L-Edit is launched provides for npn transistors using the following layering scheme:

- **Emitter**. This is ndiff = (ACTIVE) AND (NSELECT), the same as the source/drain for an nMOSFET;

- **Base**. A special layer named PBASE is provided in the process; and,

- **Collector**. The NWELL is used as the n-type collector.

Although the SCNA process is not optimized for BiCMOS operation, it does provide the basic features necessary. Metal1 connections are made using ACTIVE CONTACT and ACTIVE area combinations.

Figure 5.36 shows an example of an npn transistor created with L-Edit; the cross-sectional viewer shows the layering of the device. Performing an Extract operation on the layout yields an output of the following form:

* Circuit Extracted by Tanner Research's L-Edit V5.09 ;
* TDB File File0, Cell Cell0, Extract Definition File morbn20.ext ;

.MODEL NMOS
.MODEL PMOS
.MODEL poly2NMOS
.MODEL poly2PMOS
.MODEL NPN
Q1 9 2 8 6 NPN
* Q1 Collector Base Emitter Substrate (11 12 39 42) A = 840
* Total Nodes: 5 ;
* Total Elements: 1 ;
* Extract Elapsed Time: 21 seconds ;
.END

A .MODEL statement must be provided for the NPN transistor before it can be

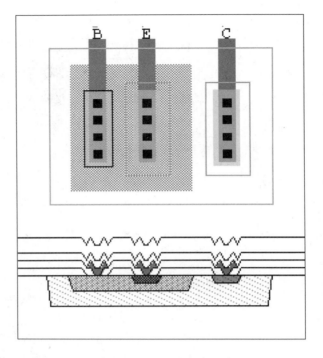

Figure 5.36. Bipolar transistor in the L-Edit SCNA technology.

used in a SPICE simulation.

5.8.3 BiCMOS Circuits

The philosophy used in constructing a BiCMOS logic circuit is to use the MOS-FETs to implement high-density logic, while providing bipolar transistors as output drivers. BiCMOS circuits are used to drive large capacitance loads, such as output pads, clock distribution networks, and long data bus extensions. Several design styles have been developed, and the reader is referred to references for in-depth treatments. Only a few circuits will be presented here.

Figure 5.37 shows a basic BiCMOS inverter. The logic function is due to MOS-FETs Mn and Mp, while bipolar transistors Q1 and Q2 are used as output drivers. The extra MOSFETs M1 and M2 are used to aid in switching Q1 and Q2 off by providing switched paths to remove base charge.

The circuit operation is best explained by examining the effect of input transitions. Suppose that initially the circuit is set with $V_{in} = 0$ v. Mn is in cutoff (OFF), while Mp is ON (active). The base of Q1 is high, so that it is biased ON. The gate of M2 is connected to the base of Q1, so M2 is also ON; however, M2 pulls the base of Q2 to ground, so Q2 is OFF. The output voltage is given by

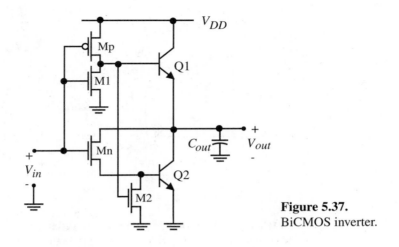

Figure 5.37.
BiCMOS inverter.

$$V_{OH} = V_{DD} - V_{BE(on)} \tag{5.33}$$

where Q1 induces a drop of $V_{BE(on)}$ from the V_{DD} level at the output of Mp. When V_{in} is switched toward V_{DD}, the operation is reversed. In this case, Mn and M1 are ON, while Mp and M2 are OFF. Q1 is OFF since M1 provides a conducting path from the base to ground. M2 is OFF, but Mn is ON. Noting that Mn is connected between the collector and base of Q2, the output voltage is given by

$$V_{OL} = V_{BE(on)} , \tag{5.34}$$

where the $V_{BE(on)}$ constraint is due to Q2. The output logic swing for this circuit is thus given by

$$V_{OH} - V_{OL} \approx V_{DD} - 2V_{BE(on)}. \tag{5.35}$$

The reduced voltage swing may cause a problem in a high-density design.

An alternate circuit with an improved logic swing is shown in Figure 5.38. In this circuit, the pull-down bipolar output driver Q2 has been replaced by a single large nMOSFET that provides logic and is used to drive the output load. The analysis for V_{OH} is the same as in the previous circuit. However, the output low voltage is reduced to $V_{OL} = 0v$, giving the output logic swing as

$$V_{OH} - V_{OL} \approx V_{DD} - V_{BE(on)}. \tag{5.36}$$

The trade-off is in the layout area. Since Mn must sink a large amount of current to discharge C_{out}, it must be very large compared to a standard logic FET such as M1.

As in other CMOS logic families, the inverter circuit can be used as a basis for series-parallel logic gates. A BiCMOS NAND2 gate is shown in Figure 5.39. The circuit uses MOSFETs MpA, MpB, MnA, and MnB to implement the NAND function. M1A and M1B are used to remove charge from the base of Q1, while M2 provides the same function for Q2. Pull-down networks of this type help to overcome

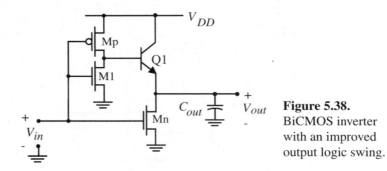

Figure 5.38. BiCMOS inverter with an improved output logic swing.

the charge storage delays associated with bipolar devices. Variations such as the NOR gate or general AOI logic follow the same structuring.

5.9 Chapter Summary

In this chapter, we have examined many of the important circuits used in digital CMOS logic design. The most important points to remember are:

- CMOS logic gates can be static or dynamic;
- The placement of the MOSFETs in the gate determines the logic function;

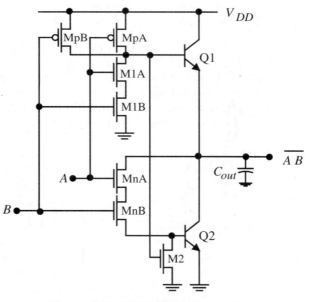

Figure 5.39. BiCMOS NAND gate.

- Electrical performance is set by the parasitic resistance and capacitance of the transistors; and,
- The layout sets all of the MOSFET aspect ratios in the circuit, which directly affects the switching characteristics.

It is important to understand the relationship between the specifics of the layout and the resulting performance of the circuit.

5.10 References

R5.1 D.A. Hodges and H.G. Jackson, **Analysis and Design of Digital Integrated Circuits**, 2*nd* ed., McGraw-Hill, 1988.

R5.2 H. Haznedar, **Digital Microelectronics**, Benjamin-Cummings, Redwood City, CA, 1991.

R5.3 J.P. Uyemura, **Circuit Design for CMOS VLSI**, Kluwer Academic Publishers, Norwell, MA, 1992.

R5.4 J.P. Uyemura, **Fundamentals of MOS Digital Integrated Circuits**, Addison-Wesley, Reading, MA, 1988.

R5.5 N.H. Weste and K. Eshraghian, **Principles of CMOS VLSI Design**, 2*nd* ed., Addison-Wesley, Reading, MA, 1993.

5.11 Exercises

In all problems, use $L=2\mu m$ (where $1\lambda= 1\mu m$) and assume that the channel length L is the same as the drawn value L'.

E5.1 Draw the layout for an inverter that is contained within a rectangular box with dimensions 50λ high and 20λ wide as shown in Figure P5.1. Use the default SCNA technology, and run a DRC on your circuit to ensure that no rules are violated. Try to use the largest transistors possible for a symmetric design, i.e., one where $\beta_n=\beta_p$.
  Run the Extract operation to verify the dimensions of the transistors in the final design.
  Finally, use SPICE to obtain the DC voltage transfer curve for the circuit.

E5.2 Use L-Edit to perform the basic layout and extraction of a static CMOS

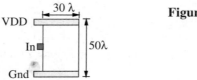

Figure P5.1

inverter circuit that has $(W/L)_n=20$ and $(W/L)_p=20$. Use the SCNA technology base.

(a) Add the MOSFET SCNA process parameters from Chapter 4, and then run a DC SPICE simulation to generate the DC voltage transfer curve for the circuit.

(b) Cascade three identical inverters together and then perform a transient simulation on the network using SPICE. Use a PULSE input to the first inverter, and examine the input and output of the middle inverter. Determine the values of t_{HL} and t_{LH} for the design. As a starting point, use a source of the form

$$\text{VSOURCE N+ N- PULSE(0 5 0 10PS 10PS 5NS 10NS)}$$

and adjust the PULSE parameters as needed.

E5.3 Construct the layout for 3 series-connected nFETs that have the same aspect ratio of $(W/L) =20$. Extract the SPICE file and verify the circuit.

E5.4 Create the layouts for the MOSFET combinations shown in Figure P5.2(a), (b), (c), and (d) with aspect ratios of $(W/L)_n=10$ and $(W/L)_p=14$ for the designs. Verify your layout using Extract on each circuit.

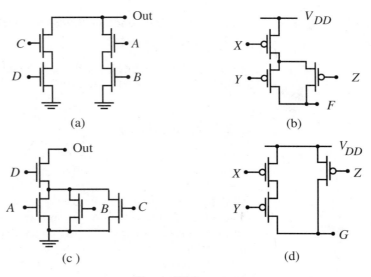

Figure P5.2.

E5.5 Design an AOI static CMOS logic gate that implements the function

$$F = \overline{AB+AC} . \qquad (5.37)$$

Start with a stick diagram, and then translate the topology to L-Edit. Choose the nFETs to have aspect ratios of $(W/L)_n=8$, and the pMOSFETS with aspect ratios of $(W/L)_p=20$. Verify the transistor connections from the Extract output.

Chapter 6

Concepts for VLSI Chip Design

The previous chapter examined the structure and characteristics of basic CMOS logic gates. However, individual gates by themselves are not of much interest. The true power of integrated system design becomes clear only after a large number of gates are wired together to form a larger system. In this chapter, we will investigate some of the basic concepts that arise in the design and layout of large digital integrated circuits.

6.1 Building Blocks for VLSI

VLSI (very large scale integration) designs may consist of several million transistors integrated into a single die. At first sight, this level of complexity can be overwhelming. However, computer-aided design (CAD) tools and intelligent logic structuring provide the key to making the design problem tractable.

6.1.1 System Structure

Although a digital logic network may perform very complex functions, it can always be broken down into simple operations. At the lowest level, the system may be described by logic expressions using the NOT, AND, and OR functions. However, it is usually more convenient to define larger functional blocks, such as multiplexers, parallel adders, and memory arrays, which provide the basic system functions. This viewpoint can be expanded to the point where even a microprocessor is just another unit in the overall design.

This simple observation illustrates the fact that digital chip design is intrinsically structured using a hierarchial approach. Every system is made up of building blocks that provide a set of basic operations. The behavior of the system is determined by how the individual units are connected together. If we apply this concept to chip design, then complex systems become tractable, providing the basis for VLSI.

When analyzing existing integrated circuit layouts, two important points become clear.

- **Regularity**. It is possible to identify regular geometrical patterns. Most objects are either rectangles, or polygons that can be decomposed into simple rectangles.
- **Replicated Cells**. The system is constructed by using basic gates and modules, so that it is not necessary to custom design every cell. Instead, primitive cells can be replicated and modified as needed.

Examples of these properties are shown in Figure 6.1; this chip design is contained on the L-Edit disk under the file name **example.tdb**, and can be accessed using the Open command in the File window of the Menu Bar. The chip itself is an optical motion detector system that consists of an array of photodetector-signal processing cells[1]. The upper drawing (a) is a close-up (ZOOM) of a typical region of the chip, and illustrates the regular patterning of the layers. The entire chip is shown in (b). At this level, the cells are easily identified.

These simple observations provide the key to intelligent design. By duplicating basic logic gates and/or logic modules, large-scale systems become tractable. Geometric regularity helps solve the problems of fitting the modules together, and providing interconnect as needed.

6.1.2 Design Philosophies

Digital VLSI can be implemented at several levels, depending upon the starting point. The most common divisions are as follows:

1. **Full Custom**. In full custom design, every detail of the integrated circuit layout needs to be completed. At this level, all gates must be designed, drawn, and simulated.

2. **Cell Based**. Cell-based designs are based on exiting cells stored in a **library**, which is a collection of pre-designed gates and modules. The properties of each cell, such as the speed and layout dimensions, are provided to the system designer, who provides the arrangement and interconnect to implement the system. Application-specific integrated circuits (ASICs) are usually constructed in this manner.

3. **Gate Arrays**. Gate arrays consist of arrays of MOSFETs that can be wired using interconnect lines to implement the desired functions. Logic circuits can be prototyped very quickly using this approach.

This list is arranged in increasing order of regularity. In full custom design, most (or all) of the circuits are designed "from scratch," without importing pre-designed cells. Cell-based design provides the basic functions, and cells can be used by themselves or as building blocks for more complex modules. Gate arrays and

---

[1] This cell was designed by John Tanner, President of Tanner Research, Inc., for his Ph.D. dissertation research at Caltech. For more information, please see the references at the end of the chapter.

related styles have pre-defined layouts, with only the interconnect patterns and contacts remaining to be specified. Other variations exist, such as Field-Programmable Gate Arrays (FPGAs), but most can be included in the above groupings.

In the treatment here, we will examine chip design from a viewpoint somewhere between the full-custom and a cell-based approach. Emphasis will be placed on creating and duplicating logic cells as the basis of chip design. However, custom circuits are often necessary due to special circumstances, and these will be discussed for a few sample cases.

6.1.3 Cells and Hierarchy

As a starting point, let us define a **cell** as a basic unit in our design. At the logic

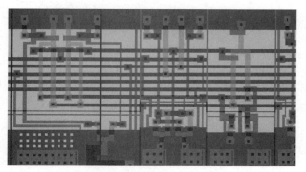

(a) Close-up view.

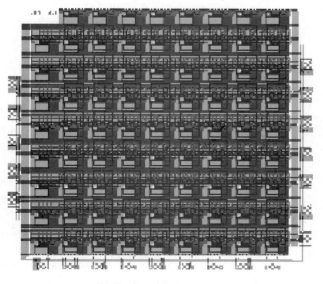

(b) Entire chip layout.

Figure 6.1. L-Edit chip example.

level, this may be a simple logic function, or a complex Boolean operation. The circuit equivalent of a cell is a defined layout pattern with given dimensions, input and output ports, and specified electrical performance. Although these are convenient for the present discussion, other definitions of a cell may be useful. For example, a cell may be a set of HDL declarations or timing diagrams.

A system is constructed by interconnecting cells together. At the chip design level, the most important factors are

- **Area and Dimensions**. Every cell consumes chip area[1] and has a particular geometrical shape associated with it. Both are important for high-density integration.

- **Ports**. The location of input and output ports is very important for routing the data path. Also, the power supply and ground are usually needed to provide electrical energy to the cell.

- **Interconnect Strategy**. The cells must be wired together using the interconnect layers. Even with three or four separate conducting layers, this can be the limiting factor in the logic density.

Cell concepts are powerful tools for designing complex integrated circuits.

As a simple example, consider a full adder circuit. This takes inputs A, B, and C_{in}, and gives outputs of

$$S = A \oplus B \oplus C_{in}$$
$$C_{ou} = AB + C_{in}(A \oplus B) \tag{2.1}$$

for the sum and carry bit, S and C_{out}, respectively. The logic diagram in Figure 6.2 illustrates how the XOR, OR, and AND gates are used to implement the function. From the cell viewpoint, only three primitive units (the three gates) are needed.

Once a full adder has been defined, then we can use it as another distinct cell

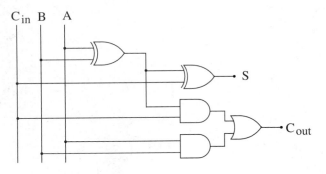

Figure 6.2. Full-adder logic.

---

[1] In the jargon of chip design, the area is often referred to as "real estate".

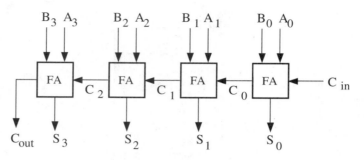

Figure 6.3. Parallel 4-bit adder diagram.

with a complex logic function. The simple ripple-carry parallel adder in Figure 6.3 shows how four full adder cells can be wired to add two 4-bit words, giving the output word (S_3 S_2 S_1 S_0) and the carry-out bit. We can raise our viewpoint another level of hierarchy by viewing the 4-bit parallel adder as yet another cell. Thus, a cell is a basic building block, regardless of the complexity of the interior. Of course, at the integrated circuit level, complex circuits require more area than simple ones, and usually exhibit longer switching times.

6.2 Using Cells in L-Edit

L-Edit provides very powerful commands for creating, editing, and applying cells in layout drawings. These are accessed from the **Cell** window of the Menu Bar. The language used to describe cells can be confusing at first. It is therefore useful to describe how cells are created and used before describing the commands in detail.

Every drawing in L-Edit is labeled with a file name, with the default name being file0. In addition, a cell name is assigned to the drawing on the screen; the default cell name is cell0. This information is in the upper left corner of the L-Edit screen, just below the Menu Bar. It is possible to create cells and store them in memory. Once a cell has been created, it can be replicated as needed in the circuit drawing. The New command is used to create a new cell. After supplying a cell name, everything you draw will be contained within the cell. Alternately, an existing cell can be accessed using the Open command. When you finish editing a cell, there are three options:

- The edited cell can be Saved by simply opening another cell. This feature is automatic; it is not necessary to execute any special command; or,

- The cell can be Renamed using the command in the Cell window; or,

- Using the Close As command keeps the original cell intact, but saves the new cell with editing changes under a new name.

The Revert command is the same as "Undo", and is used to reverse editing steps.

Stored cells can be accessed using the Instance command. When this command is executed, a list of currently available cells is displayed. This list constitutes the library for the file; any cell in the library can be copied in the layout editor. Note

---

Using Cells in L-Edit   **6-5**

```
┌─────────────────────┐
│ Cell                │
├─────────────────────┤
│ Info...             │
│ ─ ─ ─ ─ ─ ─ ─ ─ ─ ─ │
│ New...           N  │
│ Open...          O  │
│ Revert Cell...      │
│ Close As...         │
│ Delete...        B  │
│ Rename...        T  │
│ Instance...      I  │
│ Copy...          C  │
│ Fabricate...        │
│ ─ ─ ─ ─ ─ ─ ─ ─ ─ ─ │
│ Flatten             │
└─────────────────────┘
```

that the current cell (that is shown in the work area) is not available, since it has not been stored as a separate cell entity. Selecting a cell places it into the layout, where it may be selected and moved like any geometrical object. The term "instance" is used to mean a cell that has been copied into the drawing in this manner. The Delete command allows you to remove a cell in the library. A cell may be instanced as many times a needed. Alternately, a duplicate cell can be made using the Copy command and working through the dialog window; the new cell must be renamed to avoid conflicts in references. The Fabricate command is only used when preparing the layout for the fabrication process, and is not accessed during cell editing.

Cell Hierarchies

A visual representation is useful for understanding the concept of hierarchy of the cells. Figure 6.4(a) shows a chip that is made up entirely of **primitives**, i.e., that it has only basic geometrical objects such as rectangles and polygons. The example in drawing (b) shows a chip that uses both primitives and basic cells; this is meant to imply that the cells on level 1 are themselves made up of only primitives. In (c), the chip is made of cells from level 1 that may consist of just primitives, or both primitives and cells from higher levels. For example, a cell at level 1 can contain a cell from level 2, which in turn contains a cell from level 3. This type of hierarchical structuring allows you to construct cells of varying complexity, from simple gates to large-scale functional blocks.

Flattening

The last entry in the Cell window is Flatten. This command is used to reduce all cells to the level of primitives, as illustrated in Figure 6.5. The Flatten command is not reversible; once it is applied to a cell, all information concerning instanced cells is lost.

Using Cells in the Student Version of L-Edit

The Student Version of L-Edit provides all standard cell functions contained in the

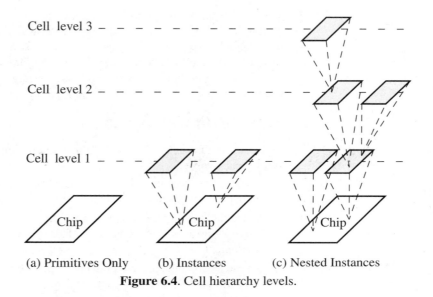

Cell level 3

Cell level 2

Cell level 1

Chip

Chip

Chip

(a) Primitives Only   (b) Instances   (c) Nested Instances

Figure 6.4. Cell hierarchy levels.

full professional version. Cells can be created, edited, and instanced as described above. However, since this version of L-Edit can only access the DOS limit of 640KB of system RAM, you may run out of memory if the cell library gets too large, or if the cells contained within the library are too complex.

Cells are stored with the current layout file, so that the Save operation copies the file and all cell libraries to a disk. The cell library will be available when the file is opened. Since the cell library is contained within the layout file, it requires RAM storage space when the file is open. This reduces the amount of working memory

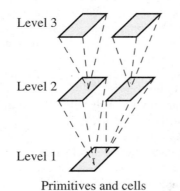

Level 3

Level 2

Level 1   Level 1

Primitives and cells   Primitives

(a) Original cell hierarchy   (b) After flattening operation

Figure 6.5. Flattening.

that is available for use in the layout drawing.

Memory usage is monitored using the Status command described in Chapter 1. As a general rule, you should occasionally check the amount of available RAM to ensure that you are not getting too close to the limit. This is particularly important when you are building a library, or are using a large set of predefined cells already stored in the file.

6.3 L-Edit Examples

Cell construction and usage in L-Edit is quite simple to master. There are three main steps involved.

1. Use the New command in the Cell window to create a new cell. Draw the cell using primitive elements (rectangles, polygons, etc.) as usual.

2. When the drawing is completed, save the cell to the library using either the Close As or the Rename command. This operation will place you in a new cell environment.

3. To use a cell in the present drawing, apply the Instance command and select a cell from the list. The cell will appear in your layout drawing, where it behaves like a single geometrical object.

The hierarchy is determined by the instances contained within the cell. To illustrate the overall approach, several cells that were created in L-Edit are shown below. The cells all have rectangular shape, and were designed with the same vertical dimension. It should be noted that individual n-wells were provided for every cell to aid in visualization. Also, n-well contacts and substrate contacts were omitted to keep the drawings simple. In practice, these contacts are mandatory, and must be included throughout the circuit.

Figure 6.6 provides examples of CMOS logic gates that were created in L-Edit. These are (a) a single inverter, (b) a double inverter, (c) a NOR2 gate, and (d) a NAND2 gate. The power supply (V_{DD}) is a horizontal METAL1 line at the top of each cell, while the ground is provided by a similar METAL1 pattern at the bottom of each cell. Inputs and outputs can be accessed at both the top and bottom of the cells using POLY connections. This composite drawing was created by using Instance to copy each cell into the current layout. At this level, it is not possible to make any changes in the contents of the cells. Instead, the original cell must be opened and edited as described above. The next time the cell is instanced, it will contain the modifications.

A CMOS transmission gate cell is shown in Figure 6.7. The cell consists of the TG itself (parallel-connected nMOS and pMOS transistors) in the right central part of the layout, and an inverter driver circuit on the left side, that has been connected to provide complementary control signals to the gates of the transistors. The Control port is used to turn the transmission gate transistors ON and OFF.

The final example is for the NOR-base SR latch circuit shown in Figure 6.8. The corresponding cell consists of left-side and right-side NOR2 gates, with the cross-coupling achieved using POLY and METAL1 interconnects and POLY CONTACT opening. This circuit provides the set operation by pulsing S to 1, and resets by

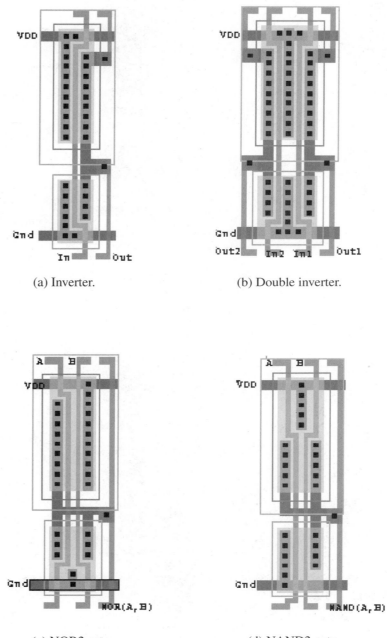

(a) Inverter.

(b) Double inverter.

(c) NOR2 gate.

(d) NAND2 gate.

Figure 6.6. L-Edit cell examples.

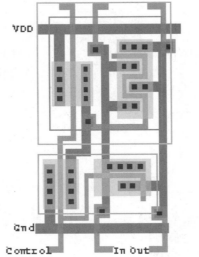

Figure 6.7.
Transmission gate layout.

pulsing R to 1. When both $S=0$ and $R=0$, the latch is in a hold state. As usual, the case where both S and R are 1's is not used.

When a cell is instanced, it initially appears in the layout drawing with an emphasized outline of the physical borders, and its name displayed. The primitives (i.e., the actual layout) are also shown. In the design of large systems, it is often desirable to view the cell as a functional block. This can be accomplished using the Show/Hide All Insides commands within the View window of the Menu Bar[1]. If the primitives are shown and you invoke Hide All Insides, only the cell box outlines

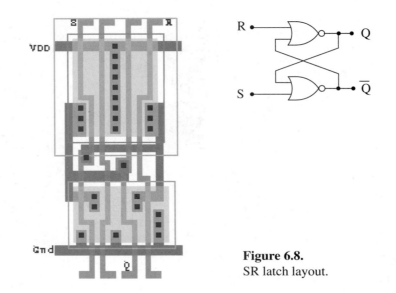

Figure 6.8.
SR latch layout.

Concepts for VLSI Chip Design

will remain; all primitives at the current cell level will remain. Figure 6.9 shows the result of this action. To restore the interior views, execute the Show All Insides command.

6.3.1 A Basic Adder Circuit

To illustrate the technique of constructing a hierarchy of cells, let's examine the step-by-step implementation of a basic parallel adder circuit with ripple carry. There are three main steps:

1. Construct a single full adder cell;
2. Replicate the cell to obtain the parallel circuit;
3. Save the parallel circuit as a single cell.

We will examine each step in detail to understand the process.

Function Definition

To implement a full adder cell in CMOS, we will use the AOI equations

$$\overline{C}_{out} = \overline{AB + AC_{in} + BC_{in}}$$

$$\overline{S} = \overline{\overline{C}_{out}(A + B + C_{in}) + ABC_{in}}$$

(2. 2)

which take inputs A, B, and C_{in}, and produce the sum S and carry out C_{out} bits. Note that the expression for the sum bit S uses the complement of C_{out}. The logic diagram for these equations is shown in Figure 6.10. Since both gates have AOI structuring, series-parallel construction can be used to design the logic circuits. These are shown in Figure 6.11.

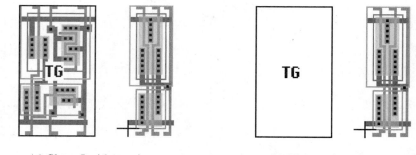

(a) Show Insides active. (b) Hide Insides active.

Figure 6.9. Using a cell instance in L-Edit.

---

[1] The command automatically changes to Show All Insides or Hide All Insides, depending on the present viewing mode.

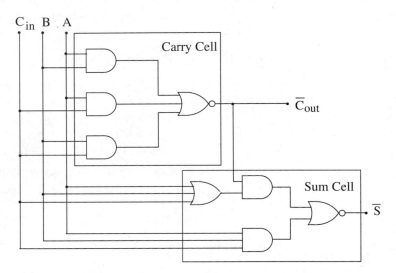

Figure 6.10. AOI logic for full-adder circuit.

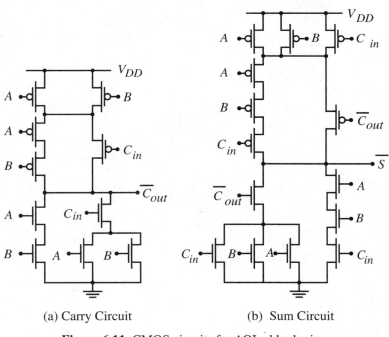

(a) Carry Circuit   (b) Sum Circuit

Figure 6.11. CMOS circuits for AOI adder logic.

Gate Layout

To illustrate the design process, we will first create separate cells for the sum and carry-out functions, and then create a composite full-adder cell. The basic cell units are shown in Figure 6.12; an inverter has been included to provide C_{out} (instead of the complement). Note that the cells have been chosen to be of equal height. In particular, the power supply and ground lines (horizontal METAL1 patterns at the top and bottom) can be matched. Also, some care has been used in routing the inputs and outputs. Bits A and B are introduced at the top of the cells. The input carry C_{in} is at the lower right side and aligns with the output from the inverter.

The three sections are then grouped to create the composite cell shown that will

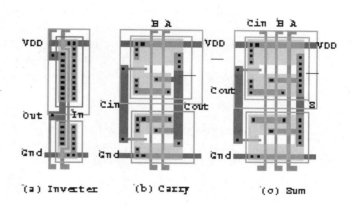

Figure 6.12. Cells used to create the full-adder.

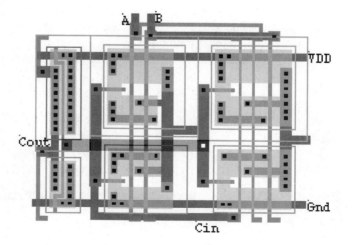

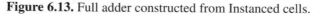

Figure 6.13. Full adder constructed from Instanced cells.

---

be named Adder. The complete cell is shown in Figure 6.13. It is seen that Adder consists of Sum, Carry, and INV cells that have been connected to provide both functions. Since the sum algorithm uses AOI logic, the Adder cell produces the complement of the sum bit. If desired, an inverter may be added to the output to regain S.

Construction of a Parallel Adder

For the next cell we will create a 4-bit parallel adder using a full-adder circuit. As shown in Figure 6.14, four individual Adder cells are connected to give parallel inputs and ripple carry. If we wish to carry the example one step farther, two 4-bit parallel adders can be paralleled to give an 8-bit adder, and so on. Note that any of the cells can be flattened if desired. This operation reduces the contents of a cell to primitive structures that can be edited directly within the cell.

This example illustrates how cells can be used to build complex chips. The important ideas that arise in the process are (a) designing basic cells that can be used to construct more complex functions; (b) employing cell geometries that can be easily interfaced with other cells; and, (c) building a hierarchy of cells.

6.4 Chip Layout

Designing large functional blocks requires a hierarchical approach. First, the individual logic gates or primitive functions are created and saved as cells. These may be used to create larger cells, which may in turn be used for even larger cells. It is important to develop a layout philosophy that provides some guidelines. Otherwise, the problem can turn into a bookkeeping nightmare!

6.4.1 Signal Groups

Digital signals may be grouped into two main categories, **data** and **control**. In general, data bits are encoded information segments, such as numbers or symbols, being processed by the system. Control bits, on the other hand, determine the operations that the data bits are subjected to. Figure 6.15 illustrates how the two act in a simple 4:1 multiplexer. The four input lines D_3, D_2, D_1, D_0, and the single output port F, transmit data bits, while the control bits ($C_1 C_0$) determine the operation,

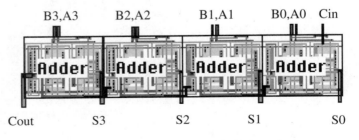

Figure 6.14. 4-bit parallel adder.

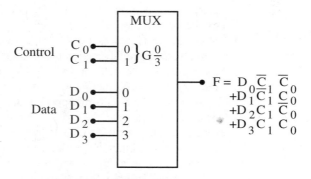

Figure 6.15. 4:1 multiplexer.

i.e., which of the four lines is directed to the output.

The design of a digital network can usually be divided into two related sub-problems: the design of the **data path logic**, and the design of the associated **data path control logic**. Designing the data path is centered on creating the functions that process the data stream. For example, logical operations, such as NANDs and NORs, or arithmetic operations, such as addition, are classified as data path functions. The control logic circuits, on the other hand, determine the route that the data follows for the current operation. Control signals are used to switch the data stream through the gates in proper order. These concepts are quite general, and can be used to describe systems of arbitrary complexity. For example, in a microprocessor, the program controls the data path sequence, while the data path itself contains the functions.

Although there are many exceptions, data path logic tends to be localized and can be contained within cell units. Control logic signals, on the other hand, are usually applied to groups of functional blocks, and must be routed to various locations. This illustrates the basis of hierarchical design. Cells provide defined logic functions with input and output ports. Once the basic functions are defined, the cells may be placed and wired as needed to provide the system level functions.

An integrated circuit represents the physical realization of a digital logic system. When implementing the design, logic gates consisting of transistors form the most basic level. The next level of design uses these gates to create logic functions and subsystems. Piecing the system together into a single chip finalizes the design process. Several important tasks must be completed to translate the system characteristics to the layout. These are common to every chip, regardless of the function.

6.4.2 The Floorplan

The **floorplan** of the chip shows the placement and area consumption of the major logic functions in the finished chip. Routing of data lines, clocks, and control signals is determined by comparing the logical design with the physical design of the floorplan.

Floorplan design is performed at the architectural level. All major operations groups are identified and the interconnect requirements are studied. The placement of the cells is not arbitrary, since the large-scale system performance is directly affected. Initial real estate allocations are based on the complexity of the cells, the limitations of the fabrication technology, and past experience. The size, shape, and placement of the cells are adjusted as needed as the design evolves. This often requires trade-offs among the various functional units to obtain a functional design.

Although the details of floorplan design are beyond the scope of this book, it is possible to illustrate the main ideas using simple concepts. In Figure 6.16, the blocks labelled Unit A, Unit B, etc., represent distinct cells, each with a specific logic function. Input and output lines are specified by port locations on the blocks. In addition, power supply and ground connections must be included. Interconnections among units (both data and control signals) must be routed within the limits imposed by the layout design rules. The lines may go around functional units, or may pass through (or over) them.

The floorplan is important to designers at all levels. At the system level, the distance between functional blocks can limit the data transmission speed and, hence, the performance of the chip. Circuit and cell design is often based on satisfying the electrical specifications using the minimum amount of real estate consumption. Interconnect routing affects every aspect of the design and operation of the network.

6.4.3 Interconnects

The limiting factor in high-density system design is the interconnect routing and connections. One reason for this situation is the existence of basic layout rules such as

- The minimum width and spacing rules for wires on the same layer, and,
- Surround design rules that are required for contacts and vias.

These automatically limit the density of the wiring. Since each layer is intrinsically two-dimensional, wires on the same layer cannot cross without creating an electri-

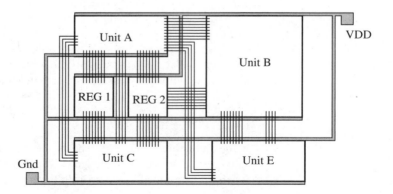

Figure 6.16. Layout concepts.

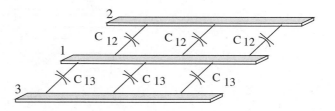

Figure 6.17. Coupling capacitance.

cal short circuit. Routing must be implemented using cross-overs and cross-unders, and the problem can become very complex. It is interesting to note that, with lithographic resolutions below 1 micron, the number of MOSFETs in a circuit is not of much concern because each transistor is so small. However, the number of contact cuts needed can be a problem due to surround-type design rule specifications.

Interconnect routing is complicated by parasitic electrical coupling among lines that are physically close to each other. This is termed **crosstalk**, and can cause data transmission errors. Consider the situation shown in Figure 6.17 where the central wire 1 is coupled to lines 2 and 3 through the distributed parasitic capacitances C_{12} and C_{13}. Crosstalk occurs because a time-varying voltage on line 2 or 3 induces an unwanted signal on line 1. To see this, first note that the charge on line 1 is given by the expression

$$Q_1 = C_1 V_1 + C_{12} V_{12} + C_{13} V_{13} , \qquad (2.3)$$

where

V_1 is the voltage on line 1 with respect to ground,
C_1 is self-capacitance of line 1 with respect to ground,
$V_{12} = V_1 - V_2$ is the voltage between lines 1 and 2, and,
$V_{13} = V_1 - V_3$ is the voltage between lines 1 and 3.

The self-capacitance C_1 is the normal interconnect-to-ground interconnect contribution. Differentiating this equation with respect to time gives the current on line 1 as

$$i_1 = C_1 \left(\frac{dV_1}{dt} \right) + C_{12} \left(\frac{dV_{12}}{dt} \right) + C_{13} \left(\frac{dV_{13}}{dt} \right) , \qquad (2.4)$$

showing that any change in voltages results in a signal on the center line. This equation also demonstrates that both positive and negative coupling can occur, depending on the polarity of the voltage differentials. In practical terms, this implies that both logic 0 and logic 1 voltage levels can be affected. This is a major reason for designing **noise margins** (using a range of voltages to represent logic 0 and logic 1 inputs) into logic gates as discussed in Chapter 5.

Crosstalk problems can be difficult to isolate, particularly in high-density layouts. It is therefore best to avoid it in the original design. The effects can be minimized by obeying all design rule spacings, avoiding long lengths of parallel lines,

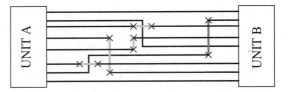

Figure 6.18. Interconnection problem.

and purposely introducing "kinks" into the lines to disturb the coupling.

Interconnect wiring usually requires that signals be routed through different layers as illustrated in Figure 6.18. In this drawing, Unit A and Unit B are to be connected by an 8-bit data bus. However, the location of the input and output ports requires that two levels of interconnect be used to achieve the desired results. When employing different material layers, it is important to recall that layers can be crossed with only minimal coupling. In the example shown in Figure 6.19, we see that

- POLY can cross METAL1 and METAL2;
- A POLY contact can be used to create a current flow path to Metal1;
- METAL1 and METAL2 can cross.

These rules help solve the interconnect problems in complex situations[1]. Note, however, that layer-to-layer coupling capacitance is introduced whenever lines cross; the value of the capacitance is proportional to the overlap area. For example, the METAL1-METAL2 overlap shown in the drawing actually couples the two horizontal METAL1 lines to each other. Although the coupling is very weak, it may cause problems in high-speed systems.

6.4.4 Padframes

Bonding pads are large metal regions that allow wiring between the silicon circuitry and the pins on the IC package. As an example, the SCNA technology uses bonding pads with dimensions of (100μm × 100 μm). The padframe layout gives the place-

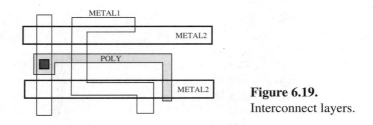

Figure 6.19.
Interconnect layers.

---

[1] An automatic Place and Route algorithm is particularly useful for wiring complex systems. Although this feature was not included in the Student Edition of L-Edit (due to the limits on memory usage), it is available in the full professional version.

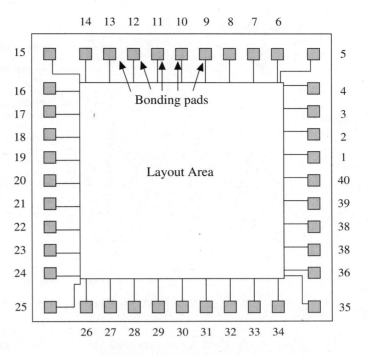

Figure 6.20. Bonding pad layout.

ment of the bonding pads and sets the maximum die size and the number of connections to the outside world. These are predefined for a given die size. Figure 6.20 shows the padframe for a 40-pin design. These have been numbered according to their pin connection when the die is placed in a dual in-line (DIP) package. Note that two bonding pads must be reserved for the power supply and ground connections.

6.4.5 Global Signal Distribution

Logic networks use both local and global signals for their operation. Local signals are most often data streams, and can be treated at the cell level. However, global broadcasts, such as system timing clocks, must be accounted for in the floorplan in order to ensure proper distribution. Interconnect network topologies are often based on geometrical symmetries in an effort to ensure that each line delivers the same signal to the receiver, regardless of the location.

6.4.6 Power Distribution

All integrated circuits require power distribution bus lines to supply current to the gates. In CMOS, we usually have a positive V_{DD} and a ground (or V_{SS}) voltage that must be routed across the die. It is important to use a regular geometrical structure that accommodates the shapes of the logic cells.

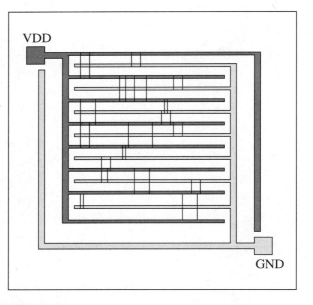

Figure 6.21. Power supply and ground distribution.

There are several geometries that are used for power supply distribution. A simple example is shown in Figure 6.21. This provides V_{DD} and *Gnd* rails with space in between for the placement of circuits. It is important to remember that power distribution lines must be capable of handling relatively large current levels. Since low resistance lines are essential, wide metal lines are used.

6.5 Input and Output Circuits

Bonding pads act as the interface between the chip and the outside world. Owing to the "harsh" external environment, special problems arise when designing the input and output circuits. It is worthwhile to briefly examine these circuits to complete the discussion.

Under normal handling of the chip, it is very easy to transfer large electrostatic voltages to the input pins of an IC package. MOSFETs are very sensitive to electrostatic discharges (ESD) of this type. If the gate voltage applied to a MOSFET exceeds a certain critical voltage V_C, the gate MOS structure will undergo a destructive breakdown, resulting in a burned-out transistor. Since even a single bad device renders the entire chip useless, this must be avoided. Input protection circuits are usually provided at input pads to prevent this from happening.

An example of an ESD protection network is shown in Figure 6.22. The objective of the network is to ensure that the CMOS inverter input V_{in} does not exceed the breakdown voltage V_{BD} that will burn out the transistors. Diodes D1 and D2 are normally in non-conducting reverse-bias states. However, if either voltage V_1 or V_2 exceeds the reverse breakdown Zener voltage V_z, then the diode goes into a con-

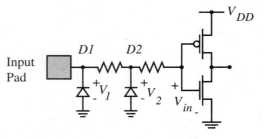

Figure 6.22. Input protection circuit.

ducting state, and allows reverse current flow to ground. This pulls charge away from the transistor gates, and insures that the voltage does not become too large. Resistors are added to further decrease the voltage that reaches the MOSFET. Implementing this type of protection network is straightforward. One approach is to use an n^+ pattern as a combination resistor/diode structure. The resistance is estimated using *t*he formula $R=R_s \times$ (Number of squares). The diode is automatically created since the n^+ region is made in a p-type background that is grounded. This is shown in Figure 6.23.

Other types of protection networks are commonly found in CMOS inputs. One popular approach uses a special FOX MOSFET built in the field oxide region. This results in a high-threshold voltage V_{TF}, typically on the order of about 12-15 volts. The circuit shown in Figure 6.24 is designed to turn on the FOX MOSFET when the input voltage exceeds VTF. When the transistor is active, it provides a current flow path to ground, keeping excessive charge away from the gate of the logic devices.

Output circuits experience a different type of problem. Since the output pad has a large capacitance, and is connected (on the outside) to a printed circuit board with a very large capacitance (on the order of pF), care must be taken to ensure that the output circuit has sufficient drive capability to maintain fast switching speeds. This requires that the output stage be designed with large aspect ratios to provide high current flow levels. However, since the input capacitance into a complementary pair of nFET and pFET devices is ($C_{Gn}+ C_{Gp}$), large transistors themselves give rise to a large input capacitance into that stage.

To avoid excessive loading (and slow down of the signal), a chain of drivers is used as shown in Figure 6.25. In this circuit, aspect ratios are designed such that

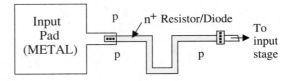

Figure 6.23. Bonding pad layout.

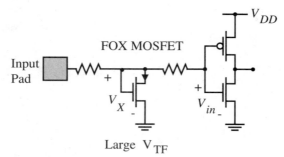

Figure 6.24. Bonding pad layout.

$(W/L)_4 > (W/L)_3 > (W/L)_3 > (W/L)_1$. The output stage has $(W/L)_4$ that is large enough to drive the large capacitance associated with the output pad and PC board. The aspect ratio $(W/L)_3$ of stage 3 is large enough to drive the input of stage 4, and so on. An analysis of this problem shows that the chain can be minimized by adjusting the aspect ratios such that every stage has the same propagation delay. For a load capacitance of C_L, and a "normal" input capacitance C_i, this requires N stages calculated from

$$N = ln(\frac{C_L}{C_i}) ; \qquad (2.5)$$

in practice, the nearest integer value is used. The scale factor is found to be approximated by the Euler $e \approx 2.71$ such that each stage increases in size by a factor $e^{(n-1)}$ for the nth stage, i.e., using the value $(W/L)_1$ as a reference, stage 2 is designed using the relation $(W/L)_2 = e (W/L)_1$, stage 3 is designed using $(W/L)_3 = e^2 (W/L)_1$, etc., until the final stage is reached. This results in the smallest (theoretical) propagation delay from the input to the output. Note that the output stage may require very large MOSFETs to meet the timing specifications. Transistors with aspect ratios of 100 or greater are not uncommon. In practice, we often start with the output stage values needed to drive C_L by analyzing the time constants. Once the out-

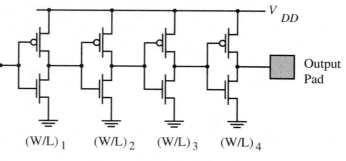

Figure 6.25. Driver chain.

put stage is designed, the aspect ratios can be scaled back toward the first stage.

6.6 Chapter Summary

In this chapter we have examined some of the fundamental concepts used to design systems in VLSI. The most important points to remember are

- Primitive functions are used to define cells that can be used to build more complex circuits;
- VLSI layout is based on structured geometries and replication of cells;
- Routing and interconnect problems become very important at the system design level.

These general principles can be applied to any chip.

This discussion concludes the tutorial material on designing CMOS integrated circuits. To read more about the subject, please consult the references listed below. However, to learn the subject of chip design, the best approach is to practice by creating layouts, simulating the circuits, and studying the interplay between the layout and the electrical performance.

6.7 References

System design of CMOS VLSI chips is a very broad subject. It is difficult to find a single reference that addresses all of the important topics. However, many books cover different aspects of the subject.

R6.1 H.B. Bakoglu, **Circuits, Interconnections, and Packaging for VLSI**, Addison-Wesley, Reading, MA, 1990.

R6.2 D.V. Heinbuch (ed.), **CMOS3 Cell Library**, Addison-Wesley, Reading, MA, 1988. This book is a complete handbook of the CMOS3 library, with each cell documented in detail. It is particularly useful for studying cell layout techniques.

R6.3 C. A. Mead, **Analog VLSI and Neural Systems**, Addison-Wesley, Reading, MA, 1989.

R6.4 S. M. Rubin, **Computer Aids for VLSI Design**, Addison-Wesley, Reading, MA, 1987.

R6.5 N. Sherwani, **Algorithns for VLSI Physical Design Automation**, Kluwer Academic Press, Norwell, MA, 1993.

R6.6 J.E. Tanner and C. A. Mead, "Optical Motion Sensor," Chapter 14 in **Analog VLSI and Neural Systems**, Addison-Wesley, Reading, MA, 1989. This chapter discusses the circuit portrayed in Figure 6.1, and available on the L-Edit disk as example.tdb.

R6.7 N.H. Weste and K. Eshraghian, **Principles of CMOS VLSI Design**, 2*nd* ed., Addison-Wesley, Reading, MA, 1993.

6.8 Exercises

Many of the exercises below require you to design basic logic cells. You may want to set a standard height (or, width, depending on how you choose the power supply and ground connections) for every cell so that the results can used in a single library. A height of 100 to 150 microns (lambda) will allow you to build reasonably complex cell functions.

Problem 1 should be completed first to gain experience in creating cells with a standard shape.

E6.1 Design an inverter cell with aspect ratios of $(W/L)_n = 10$ and $(W/L)_p = 20$ to fit in a rectangular area. Align the input and output ports so that they may be cascaded; use METAL1 for both the input and output.   Provide V_{DD} and ground lines at the

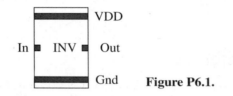

Figure P6.1.

top and bottom, so that the cell has the general structure shown in Figure P6.1. Be sure to run a DRC and an Extract operation to ensure that your design is correct.

Save the cell with the name INV. Then use the Instance command to create a new cell that consists of 6 cascaded inverters. Examine the structure of the new cell, and then use the Flatten command. Describe the differences in editing capabilities before and after the Flatten command is used.

E6.2 Create a basic cell library consisting of the following gates: INV, NAND2, NOR2. Use a rectangular cell shape and make sure that the power supply and ground wires match. Then try to create the following functions using only the cells in your library.
   (a) $F = AB$.
   (b) $G = AX + B\bar{X}$.

E6.3 Create a cell that implements the logic function f=AB+CD by using an AOI gate cascaded into an inverter. Then construct the XOR and XNOR functions. You will need to Instance the INV operation.

E6.4 Construct a cell for the logic diagram shown in Figure P6.2.

E6.5 Create a transmission gate cell that includes the INV function for gate control. Design the layout so that it can be interfaced with the static logic circuits already in

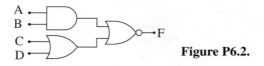

Figure P6.2.

your library. Perform a DRC and an Extract operation to verify the operation.

E6.6 A transmission gate 2:1 MUX circuit is shown in Figure P6.3. Create a cell that performs this function. Then create a 4:1 MUX using this cell as a basis.

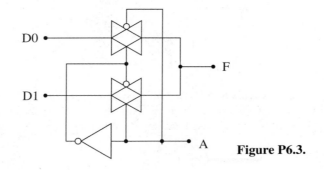

Figure P6.3.

E6.7 A simpler 2:1 MUX that only uses nMOSFETs is shown in Figure P6.4. Although this is a simpler network, it cannot pass a high voltage V_{DD} to the output under normal operating conditions.

 (a) Create a cell for this network. Be sure to include an inverter for the switching signal. Then perform a circuit Extract operation and examine the pass characteristics using SPICE.

 (b) Build a 4:1 MUX using this network. What are the strong and weak points of this design as compared to the TG-based circuit?

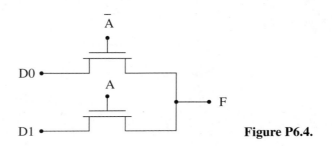

Figure P6.4.

E6.8 A simple shift register is shown in Figure P6.5. Use the TG and INV cells from your library to create a 4-bit wide register. Be careful to note the polarity of the clock signal ɸ applied to each gate.

Perform an Extract on the circuit, and use the result to perform a SPICE simulation of the circuit.

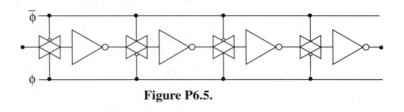

Figure P6.5.

E6.9 Create a TG-based bit-storage register cell as discussed in Section 5.6.2 of the previous chapter. Then use your design to create the layout for the 4-bit parallel data-bus transfer scheme shown in Figure P6.6.

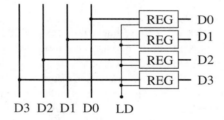

Figure P6.6.

Chapter 7

L-Edit Command Summary

This chapter is a general reference to the commands in L-Edit. Each command is listed, and a brief description of the operation is given.

7.1 Direct Keyboard Commands

Several commands can be accessed directly from the keyboard if desired. The ones listed in this section are primarily those used for navigation and drawing in L-Edit.

7.1.1 Navigation

The basic navigation commands are for PAN and ZOOM. PAN direction is controlled by the arrow keys[1] as shown in Figure 7.1. Pressing SHIFT while using the arrow keys directs you to the end of the page.

(a) PAN

(b) PAN to edge of geometry

Figure 7.1. PAN keyboard commands.

The ZOOM functions can be accessed from the keyboard as shown in Figure 7.2. The ZOOM IN magnifies the view by a factor of 2, and is used to see detail views; this is useful for constructing layouts at the transistor level. To see a larger portion of the layout, use ZOOM OUT. Each ZOOM OUT operation reduces the magnifi-

---

[1] Pan and zoom operations are discussed in Chapter 1.

cation by a factor of (1/2). The ZOOM commands can also be accessed from the menu bar heading View.

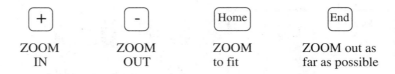

| ZOOM | ZOOM | ZOOM | ZOOM out as |
| IN | OUT | to fit | far as possible |

Figure 7.2. ZOOM keyboard commands.

7.1.2 The Home Key Command

A useful feature of L-Edit is the "home" command. This automatically sizes the entire layout to fit within the Work Area of the L-Edit screen. Simply press the [home] key on the keyboard to effect this operation. This can also be accessed from the View window of the Menu Bar, as described later.

7.1.3 Function Keys

The function keys provide a direct way to invoke the drawing tools. These are listed in Figure 7.3. Note that the drawing functions are usually chosen using the mouse.

7.1.4 Miscellaneous Commands

Several other commands can be accessed directly from the keyboard, and are summarized in Figure 7.4.

7.1.5 Redraw

A convenient feature of L-Edit is Redraw, which can be accessed by pressing the

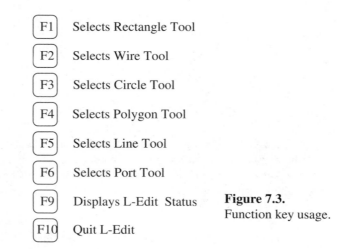

F1 Selects Rectangle Tool

F2 Selects Wire Tool

F3 Selects Circle Tool

F4 Selects Polygon Tool

F5 Selects Line Tool

F6 Selects Port Tool

F9 Displays L-Edit Status   **Figure 7.3.**
Function key usage.

F10 Quit L-Edit

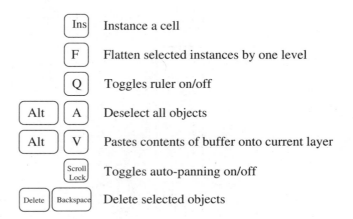

Figure 7.4. Miscellaneous keyboard commands.

space bar. Redraw erases the screen and reconstructs the layout from the information in the L-Edit buffer. This is useful when editing changes interfere with the other objects.

 Redraw screen

Figure 7.5.
Redraw command.

7.1.6 Emergency Abort

L-Edit may be aborted by using the Control-Break command:

| Ctrl | | Break | | Emergency abort

Figure 7.6. Emergency abort.

This should only be used in an emergency situation. When executed, it interrupts current program execution and returns control to L-Edit. It should be noted that internal data structures may be left in indeterminate or improper states. Once executed, it is strongly recommended that you save your work immediately, or, Quit the current L-Edit session and restart the program.

7.2 The Menu Bar

L-Edit provides access to the command groups on the Menu Bar shown in the general screen of Figure 7.7. To access a group, simply point to the group, click on the

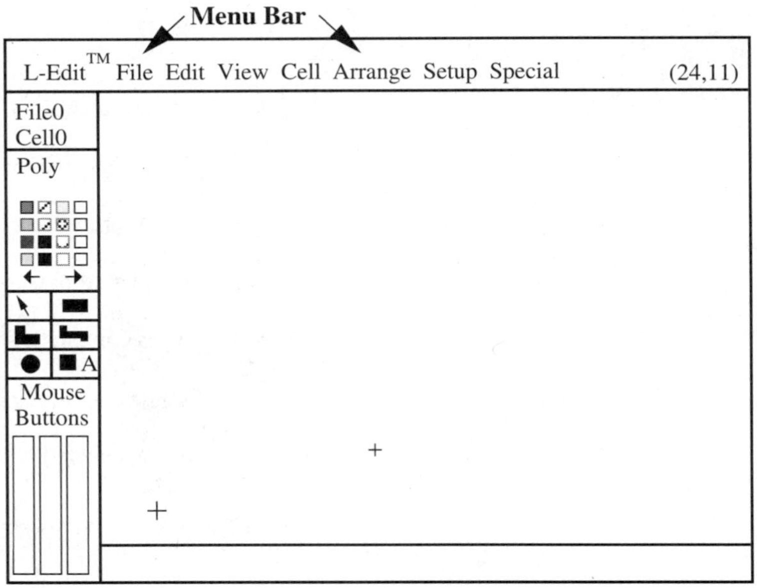

Figure 7.7. Menu Bar command listing.

left mouse button, and drag the window down. There are eight separate headings:

- L-Edit
- File
- Edit
- View
- Cell
- Arrange
- Setup
- Special

Each menu heading contains several commands. The first seven groups deal with various aspects of the program, while the Special heading allows access to the more advanced features. The student version of L-Edit contains all of the major tools provided with the full version, including circuit extraction, a design rule checker (DRC), and a cross-section viewer that allows you to see the layers in the finished chip. Every effort was made to ensure that this version of L-Edit would be useful in both classroom and research applications.

Each section in this chapter describes a separate command group. The commands are summarized by menu bar headings. Equivalent keyboard commands (if ones exist) are shown on the right side of the menu listing. The symbol "^" indicates that the Control key should be pressed at the same time as the following key, e.g., ^K means pressing Control and K simultaneously.

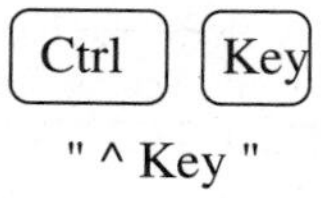

" ^ Key "

Figure 7.8.
Use of the control key.

The features available under Special are somewhat complicated, and are discussed in detail in the next chapters. Each feature is described in both general and specific terms.

7.3 L-Edit

Accessing this line in the menu bar opens the window shown in Figure 7.9. There are two choices, About L-Edit and Status.

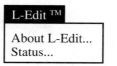

Figure 7.9.
The L-Edit window.

About L-Edit

This provides information about the L-Edit program itself including items such as the version number. In addition, the address of Tanner Research is given if you would like to have more information on their products.

Status

The Status command is your "window" to memory usage. This version of L-Edit is restricted to the 640K RAM limit imposed by DOS. Choosing this line results in a table that summarizes important quantities, such as available memory, total used, and amount of free memory.

A typical status screen in shown in Figure 7.10. There are four columns shown.

- **Memory Status**: This column lists the major categories for memory usage.
- **Count**: The count gives the number of occurrences in the present layout.
- **Bytes**: This is the total number of bytes for each category.
- **%**: The relative percentage of memory used by each listing.

For small layouts, the memory limitation will not be apparent. However, large drawings naturally require large amount of memory, since the description of each object and its location must be stored in the editing buffer. If you start to run out of memory, L-Edit will give you a warning notice. You should save your work on a regular basis.

7.4 File

The File command group provides access to normal file operations associated with

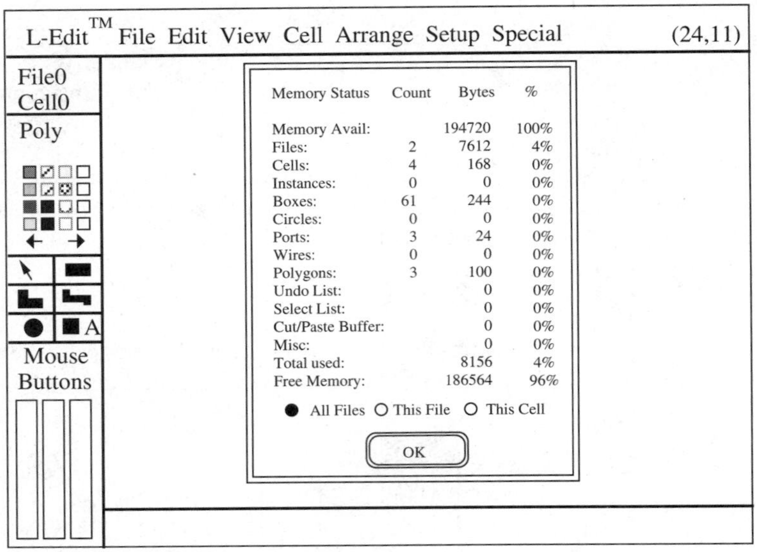

Figure 7.10. The Status box.

creating and saving your work. Opening the File heading gives the window shown in Figure 7.11.

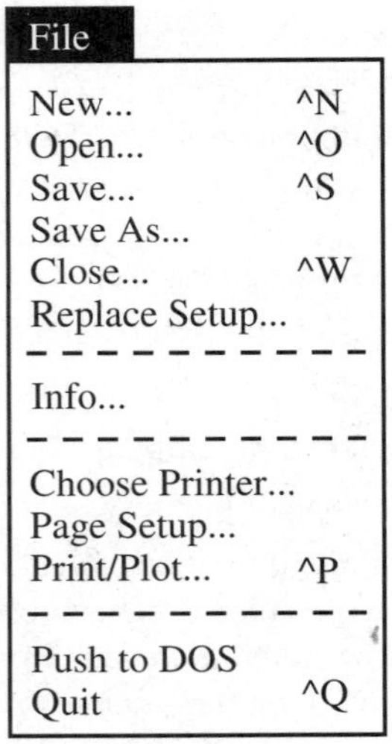

Figure 7.11
The File window.

New

This command opens a new file within L-Edit. When this command is activated, you will be prompted to enter the name of the new file, then click OK in the dialog box. The keyboard equivalent is ^N. Any subsequent Save command (see below) will automatically write to the current file.

Open

Opens an existing file from a disk. The student version can only open files with .tdb extensions. When a file is opened, you must provide the DOS path and the name of the file in the dialog box, and then click OK. This command can be accessed from the keyboard using ^O (Control-O).

Save

This is used to save the current layout in the L-Edit buffer to a disk. The file is always saved in .tdb form. If the default name File0 is currently displayed, you will be prompted to enter a name. From the keyboard, use ^S to initiate this command.

Save As

The Save As command is similar to Save, except that you may save the current work under a new file name. You must provide the new file name and DOS path when executing this command.

Close

Used to close the current file **without** writing the contents to the disk. L-Edit will warn you that your work will not be saved, and you must answer the prompt box before the file closing is completed. The keyboard combination ^W may be used as an alternate.

Replace Setup

This command allows you to modify the information in the current file by replacing it with data from a different design file. When this command is accessed, a dialog box appears. You will be prompted to enter information on the new setup parameters.

Info

Info gives information about the current design file. This includes items such as the file name, the author (if specified), the file creation date, the last date of modification, messages, and version numbers.

This command can also be used to Lock the file. A locked file can be viewed or copied, but cannot be changed.

Choose Printer

When this command is activated, you will be presented with a list of printers that can be used with this version of L-Edit. Included are several Epson-brand formats, Postscript, and plotter outputs. Selecting a printer type installs the proper drivers into the print spooler.

Page Setup

Page Setup allows you to control the page layout and the print quality when your file is sent to the printer.

Print/Plot

Choosing Print/Plot initiates the file for printing. A dialog box allows you to select various options such as the page scaling, the print range, and the number of copies.

Push to DOS

This provides a temporary escape to DOS from L-Edit. Invoking this command replaces the L-Edit screen with the standard DOS prompt, and DOS instructions may be entered. This requires at least 200K of free memory to execute. To return to L-Edit, just type EXIT.

It is important to note that you will be working in a DOS shell within L-Edit. Do not install any memory-resident programs, since they may interfere with the operation of the program when you return.

Quit

The Quit command ends execution of L-Edit and returns you to DOS. If you have unsaved work in the L-Edit buffer, a warning will be issued before the program quits. You must respond to the prompt to leave the program. The keyboard equivalent is ^Q (Control-Q).

7.5 Edit

Editing commands are used to manipulate geometrical objects in the layout. The Edit Menu is shown in Figure 7.12.

Undo

This reverses the action of the previous command and restores the layout to the form before the command was given. This command only works with commands that directly affect objects: Draw, Copy, Edit, Move, Instance, Group, Flip, Rotate, Cutting, and Merging. Each use of Undo affects the commands in reverse order, allowing you to Undo a sequence of editing steps. The keyboard equivalent is ^Z.

Cut

When the Cut command is used, all currently selected objects are removed from the layout and placed into the past buffer. To select a group of objects, use the Arrow Tool and surround the desired set; an object must be completely contained within the box to be included.The Paste command is used to move the objects back to the layout. The keyboard command ^X may also be used to cut an object.

Copy

All selected objects are copied to the paste buffer while still remaining in the lay-

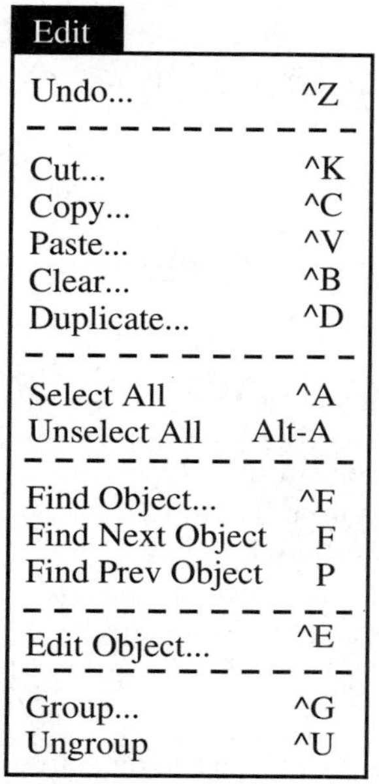

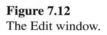

Edit	
Undo...	^Z
Cut...	^K
Copy...	^C
Paste...	^V
Clear...	^B
Duplicate...	^D
Select All	^A
Unselect All	Alt-A
Find Object...	^F
Find Next Object	F
Find Prev Object	P
Edit Object...	^E
Group...	^G
Ungroup	^U

Figure 7.12
The Edit window.

out. To select a group of objects, use the Arrow Tool and surround the desired set; an object must be completely contained within the box to be included. The keyboard equivalent is ^C.

Paste

Objects in the Paste buffer are moved back to the layout screen. Once they are back on the screen, they can be moved and manipulated as regular objects. Note that the object still remains in the buffer, and can be Pasted again until it is replaced by a Copy command. The keyboard command ^V can also be used to invoke the Paste operation.

Clear

The Clear command deletes all selected objects in the layout without copying them into the Paste buffer. From the keyboard, use ^B, or Backspace, or Delete to eliminate selected objects.

Duplicate

The duplicate command creates a copy of the selected objects. The copy appears one grid point away from the original objects, and it can be moved to the desired position using the mouse drag action. If the Duplicate command is accessed immediately after the new object is placed, the next copy will be spaced at the same relative distance. This is very useful in creating replicated structures such as arrays. The keyboard command ^D may be used if desired.

Select All

This selects all objects in the layout. Typing ^A from the keyboard invokes this command.

Unselect All

This unselects all objects that are currently selected. From the keyboard, use the combination Alt-A.

Find Object

This is used to find an object of a given type (or types) on a given layer. A dialog box allows you to select one of three types: Object, Port, or Instance. Only a single layer can be searched at a time. The keyboard command ^F may also be used to initiate the Find Object command.

Invoking this (or other Find...) command after the DRC has been run allows you to find the location of design rule violations.

Find Next Object

Find the next object of a given type and on a given layer as defined by the previous Find Object command.

Find Prev. Object

Returns to the object found by the last Find Object search.

Edit Object

Edit the selected object using text entries. This includes information such as the object's position and coordinates, the layer, data type, and other attributes.

Group

This command creates a new cell containing the selected objects and instances in the current cell.

Ungroup

Ungroups the selected instance in the current cell, and removes one level from the cell hierarchy.

7.6 View

View commands are used to control the drawing environment in the work area. The menu is shown in Figure 7.13.

Show/Hide All Insides

Choosing the Show All Insides command displays the details of all instanced cells

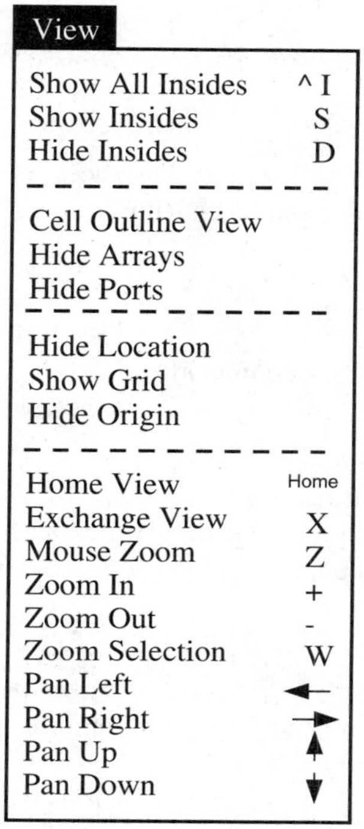

```
┌─────────────────────────┐
│ View                    │
├─────────────────────────┤
│ Show All Insides    ^ I │
│ Show Insides        S   │
│ Hide Insides        D   │
│ ─ ─ ─ ─ ─ ─ ─ ─ ─ ─ ─   │
│ Cell Outline View       │
│ Hide Arrays             │
│ Hide Ports              │
│ ─ ─ ─ ─ ─ ─ ─ ─ ─ ─     │
│ Hide Location           │
│ Show Grid               │
│ Hide Origin             │
│ ─ ─ ─ ─ ─ ─ ─ ─ ─ ─     │
│ Home View         Home  │
│ Exchange View       X   │
│ Mouse Zoom          Z   │
│ Zoom In             +   │
│ Zoom Out            -   │
│ Zoom Selection      W   │
│ Pan Left            ←    │
│ Pan Right           →    │
│ Pan Up              ↑    │
│ Pan Down            ↓    │
└─────────────────────────┘
```

Figure 7.13
The View window.

in the layout. This command is active when the details of the cells are not being shown.

If the details of the Instanced cells are visible, then executing the Hide All Insides command (which will be displayed in the Menu) will result in simple boxes with cell names being displayed. In this mode, the contents of the cell instances will not be seen.

Show Insides

This shows the interior details of the currently selected Instanced cells.

Hide Insides

The Hide Insides command removes the interior details from a cell drawing, leaving only the outline with a cell name for viewing.

Cell Outline/Icon View

With the Cell Outline or Cell Icon command, you can show Instanced cells using the information drawn on the Icon Layer. For example, an inverter cell can be described by an equivalent logic symbol on the Icon layer. This is useful for tracing logic flow in a network.

Hide/Show Arrays

This command hides (or Shows) arrayed instances of a cell.

Hide/Show Ports

Hide Ports removes port locations on instances of the cell currently being edited. Show Ports is the opposite, and labels all port locations.

Hide/Show Location

The mouse pointer location coordinates on the right side of the Menu Bar are suppressed (or shown) with this command.

Show/Hide Grid

The Show Grid command is used to give visual location of the grid points. If you execute Hide Grid, the points are not explicitly shown. Note that a ZOOM OUT operation drops the grid points from the screen when they get too dense.

Hide/Show Origin

This suppresses the origin (0,0) crosshair marker on the drawing. The origin is useful for finding DRC errors and for maintaining relative locations of cells.

Home View

The Home View command automatically scales and repositions the layout so that all objects fall within the viewing area of the screen. This is very useful for scaling the layout to exactly fit into a single viewing area.

Exchange View

Exchanges the current view with the previous view.

Mouse Zoom

The mouse is enabled for one zoom or pan operation.

Zoom In

Increases the magnification of the work area by a factor of 2.

Zoom Out

Decreases the magnification of the work area by a factor of 2.

Zoom Selection

This command zooms to include the selected objects.

Pan Left

Moves the viewing window to the left.

Pan Right

Moves the viewing window to the right.

Pan Up

Moves the viewing window up.

Pan Down

Moves the viewing window down.

7.7 Cell

Cells are primitive circuits that can be duplicated and used in other parts of the design. The Cell menu window is shown in Figure 7.14. These operations are useful for constructing large systems from basic units.

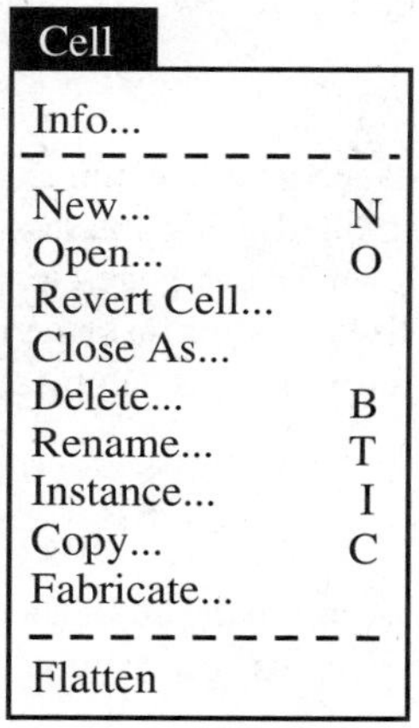

```
┌─────────────────────┐
│ Cell                │
├─────────────────────┤
│ Info...             │
│ ─ ─ ─ ─ ─ ─ ─ ─ ─ ─ │
│ New...          N   │
│ Open...         O   │
│ Revert Cell...      │
│ Close As...         │
│ Delete...       B   │
│ Rename...       T   │
│ Instance...     I   │
│ Copy...         C   │
│ Fabricate...        │
│ ─ ─ ─ ─ ─ ─ ─ ─ ─ ─ │
│ Flatten             │
└─────────────────────┘
```

Figure 7.14
The Cell window.

Info

The Info command provides a description of the current cell, such as the filename and creation dates. Cells can be locked or unlocked from this window.

New

This command is used to open a new cell. When this command is executed, the current cell is closed, but still resides in memory.

Open

The Open command is used to open an existing cell. A list of available cells is displayed when this command is executed. Choosing the desired cell name brings it into the editing buffer. The previous cell is saved in memory for future access.

Revert Cell

Revert Cell is used to restore the cell to the form it was in when first opened. All changes made since the cell was opened are nullified.

Close As

This allows you to save the cell under a different cell name and close the cell with the new name.

Delete

The Delete command is used to delete a cell from the current list. When this is executed, a list of cells will appear in a dialog window. Selecting a cell deletes it from the list.

Rename

Renames the currently open cell. The new cell name is entered in the dialog box.

Instance

This command is used to instance a cell. When executed, a list of available cells is displayed. The list may be scrolled using the arrows in the list, or from the keyboard using the Up Arrow, Down Arrow, Page Up, and Page Down keys.

Selecting a cell from the list gives an Instance of the cell in the current layout. The Instanced cell may be manipulated as an object, but you cannot edit the interior of the cell unless you Open it.

Copy

Creates a copy of a cell. This command replicates the selected cell, including all instances and primitives, and copies to a new cell. You must provide a new cell name.

Fabricate

Tag a cell for fabrication only. This command is not used except when a design is being prepared for submission to a foundry for fabrication.

Flatten

This command flattens the hierarchy in the current cell. All Instanced cells are reduced to primitives.

7.8 Arrange

Arrange commands allow you to perform basic rotations, flips, and cuts. The Arrange menu is shown in Figure 7.15.

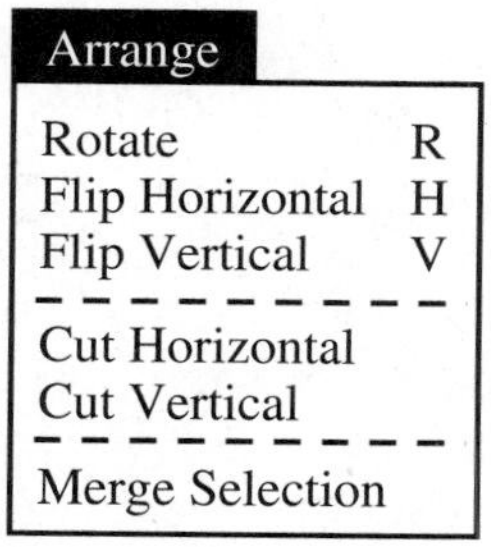

```
┌─────────────────────┐
│ Arrange             │
├─────────────────────┤
│ Rotate          R   │
│ Flip Horizontal H   │
│ Flip Vertical   V   │
│ ─ ─ ─ ─ ─ ─ ─ ─ ─   │
│ Cut Horizontal      │
│ Cut Vertical        │
│ ─ ─ ─ ─ ─ ─ ─ ─ ─   │
│ Merge Selection     │
└─────────────────────┘
```

Figure 7.15
The Arrange window.

Rotate

This rotates a cell or instance by 90 degrees. The object must first be selected for this operation, and all other Arrange operations, to be active.

Flip Horizontal

Used to flip an object or instance horizontally, i.e., around an imaginary vertical axis.

Flip Vertical

This command flips an object or instance vertically, i.e., around an imaginary horizontal axis.

Cut Horizontal

Divides the selected object(s) through a horizontal axis.

Cut Vertical

Divides the selected object(s) through a vertical axis.

Merge Selection

Merges selected objects together.

7.9 Setup

The Setup commands are used to provide specific information on the technology base. This includes material properties, such as resistance and capacitance, grid definition, design rule values, palette colors, and other parameters. The Setup menu, shown in Figure 7.16, provides the interface commands needed to specify the details of each layer.

Palette

Sets the composition of the color palette. This command allows you to modify the colors that make up the palette using a point-and-click dialog window.

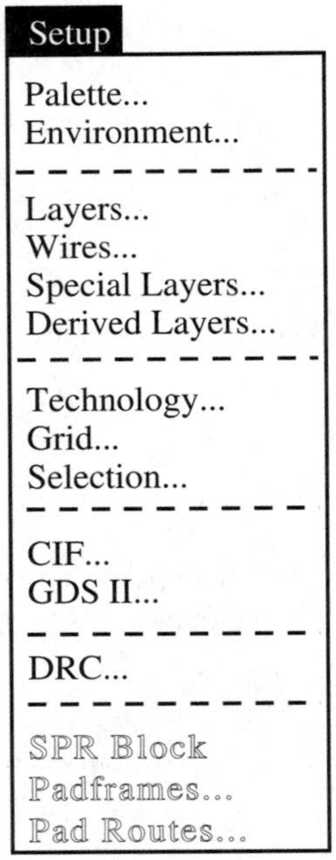

```
┌─────────────────┐
│ Setup           │
├─────────────────┤
│ Palette...      │
│ Environment...  │
│ ─ ─ ─ ─ ─ ─ ─ ─ │
│ Layers...       │
│ Wires...        │
│ Special Layers...│
│ Derived Layers...│
│ ─ ─ ─ ─ ─ ─ ─ ─ │
│ Technology...   │
│ Grid...         │
│ Selection...    │
│ ─ ─ ─ ─ ─ ─ ─ ─ │
│ CIF...          │
│ GDS II...       │
│ ─ ─ ─ ─ ─ ─ ─ ─ │
│ DRC...          │
│ ─ ─ ─ ─ ─ ─ ─ ─ │
│ SPR Block       │
│ Padframes...    │
│ Pad Routes...   │
└─────────────────┘
```

Figure 7.16
The Setup window.

Environment

Used to set miscellaneous environmental items such as highlighting, foreground color, background color, and other features of L-Edit.

Layers

Edits the layer structure of the current file. This allows you to change the way each layer is rendered, or to change the name of the layer.

Parasitic resistance and capacitance values can be entered for each layer using the dialog window. The values may be used by L-Edit/Extract when translating the physical design into a SPICE-compatible circuit listing.

Wires

Used to set the default width and style of the wires drawn by the wire tool.

Special Layers

Designates special layers.

Derived Layers

Defines derived layers using Boolean operations. A dialog window is activated, and new layers can be defined by selecting the layers to be used and clicking on the appropriate logical operations.

Technology

Edits the technology name and scaling parameters. This is useful for checking the current technology being used in the layout.

Grid

Sets the grid parameters and provides control of the mouse-pointer movement.

Selection

Sets Selection and Edit ranges.

CIF

The CIF command allows you to modify technology layers to conform with legal CIF syntax. **This feature is not available in the Student Edition of L-Edit.**

GDS II

Sets up GDS II reading and writing information. **This feature is not available in the Student Edition of L-Edit.**

SPR Block

This command is used in the automatic router in the full version of L-Edit. **This feature is not available in the Student Edition.**

Padframes

This command is used in the automatic router in the full version of L-Edit. **This feature is not available in the Student Edition.**

Pad Routes

This command is used in the automatically router in the full version of L-Edit. **This feature is not available in the Student Edition.**

7.10 Special

The commands under this heading provide access to the advanced features provided with L-Edit. These are listed on the menu shown in Figure 7.17. The remaining chapters of the book are devoted to more detailed discussions of each function.

Generate Layers

This command is used to generate derived layers. It may be used to define physical layers using the effect of two or more masks.

Clear Gen'ed Layers

Deletes all objects on derived layers.

Special
Generate Layers...
Clear Gen'ed Layers...

DRC...
DRC Box...
Clear Error Layer...

Place and Route...

Extract...

Cross-Section...

Figure 7.17
The Special window.

DRC

Activates the design rule checker algorithm within the L-Edit environment. This command runs a DRC on the entire layout. The DRC algorithm checks the layout for design rule violations, and provides both a graphical and a text file output.

DRC Box

Used to run a DRC on a defined region of the chip. The mouse is used to outline the layout area to be checked. It allows the user to perform a DRC on a small area instead of having to analyze an entire chip.

Clear Error Layer

Deletes all objects on the Error Layer. The Error Layer is generated when the DRC is run, and provides the location of design rule violations.

Place and Route

This command executes the Place and Route option. **This feature is not available in the Student Version of L-Edit**.

Extract

Runs the circuit extraction routine that translates the layout into a SPICE-compatible text file. Device definitions are provided in files with .ext extensions. For example, the file morbn20.ext is the extract definition for the SCNA process.

Cross-Section

The Cross-Sectional viewer draws a side-view of a horizontal cut of the layout pattern. The drawing is based on the specifics provided in a .xst file. For example, the SCNA fabrication process is described in the file morbn20.xst.

Chapter 8

Cross-Section Viewer

The cross-section viewer feature of L-Edit allows you to see the side view of the chip that is described by a layout. This feature is particularly useful for those new to CMOS chip design, as it provides the connection between device structures and layout. Many people find that the cross-sectional viewer is the missing link to understanding the characteristics of the individual layers and how they stack to form physical devices. It is also of interest for investigating more complex, multi-layer regions of the chip.

8.1 Accessing the Cross-Section Viewer

To initiate this feature, first pan the Work Area to include the region that you want to view. When the cross-section is drawn, it will have the same scale as the corresponding layout, so you may want to adjust the Zoom. Next, choose the cross-section viewer option from the Special grouping in the Menu Bar. This results in the dialog box shown in Figure 8.1. If no process name appears in the box, then you must specify the name of the file that contains the information on the layer sequence. Several files have been included on your L-Edit disk, and are identified by .xst extensions. For the default SCNA technology, the file is named morbn20.xst. Type in the name of the file and click OK. Finally, you will then be prompted to choose a horizontal line to specify the layout section region to be viewed. This initiates the calculation and drawing process. The cross-sectional drawing will appear in the lower half of the screen.

In normal mode operation, L-Edit generates the cross-sectional viewer draws in a continuous manner by growing each layer, etching where necessary, depositing oxide, etc., until the entire process is completed. Choosing the Single Step Display feature in the dialog box causes L-Edit to generate the cross section view one layer at a time. The mouse button is used to sequence through the steps.

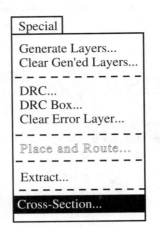

If you would like to obtain a side view of the layout, group the objects together and execute a Rotate command. The drawing may be restored rotating the objects back to their original orientation.

L-Edit does not provide direct printing of the cross-sectional views. However, a graphics screen dump program may be used to obtain hard-copy if desired. Since these are memory-resident, caution should be exercised to ensure that you do not run out of memory and lose data. Before attempting to perform a screen dump using this type of program, save your file in case a problem arises.

8.2 Process Description

The cross-sectional viewer feature of L-Edit accesses a text file for information regarding the stacking of the layers. In general, the file describes the buildup of the chip on a layer-by-layer basis using the basic fabrication steps

- **Grow/Deposit**
- **Etch**
- **Implant/Diffuse**

as needed. Each line in a process description file contains five data fields: Step, Layer Name, Depth, Label, and Comment. A space or a tab is used as a delimiter between adjacent entries. The specification for Step, Layer Name, and Depth are

Process Definition File:

morbn20.xst

☐ Single Step Display

[Cancel]   [OK]

Figure 8.1.
Cross-sectional
viewer dialog box.

described below. The Layer Name provides the connection between the file description and the layout drawing.

Grow/Deposit (gd)

This command is used to describe a material layer that exists above the surface of the silicon substrate. Oxide growths and metal depositions are described using this step, denoted by "gd" in the process description. The depth (or, thickness) must be specified in relative units so that L-Edit can provide a scaled drawing.

Etch (e)

The etch specification is used to determine which regions of a material layer are to be removed to create the desired patterns. The abbreviation "e" is used in the Step identifier. The Layer Name is given as a Boolean statement, such as NotPoly, a defined etch, such as Via, or a convenient label encased in quotes ("name"). The value of the depth indicates the thickness of the etch step.

Two comments should be made concerning the Etch command. First, all layers are etched uniformly, so that physical processes such as etch-stoppers cannot be simulated. Second, the resulting cross-sectional views are those for a reasonably isotropic etch process where the vertical and lateral (horizontal) etch rates are almost the same. This means that the sidewalls are not vertical, but will have a noticeable slant to them.

Implant/Diffuse (id)

The Implant/Diffuse step, denoted by "id" in the process description file, is used to indicate that the silicon layer nearest to the surface is doped using either an ion implantation, or a high-temperature impurity diffusion. This step is used to create n- and p-wells, in addition to n^+ and p^+ doped regions.

8.2.1 File Syntax

As mentioned above, the cross-sectional viewer constructs the chip using a step-by-step description of the fabrication sequence. In a process description file, every step is described by a line consisting of the Step, Layer Name, Depth, Label, and Comment entries. L-Edit only displays layers that are described by the file. This includes the substrate and all subsequent materials.

Let us examine a few lines to understand how each command works. First, consider the statement

Step	Layer Name	Depth	Label	Comment
gd	-	20	p-	# 1.Substrate

This command grows (gd) a lightly doped (p-) substrate with a thickness of 20 units. Note that comments are preceded by a pound sign "#"; L-Edit ignores all text to the right of the "#" delimiter.

Next, suppose that we want to create an n-well in the p-substrate. The following implant/diffuse command is used.

Step	Layer Name	Depth	Label	Comment
id	"Well X"	2	n-	# 2. n-well

The depth of the n-well is defined to be 2 units deep.

The next example uses the following steps:

1. Grow field oxide
2. Etch active areas
3. Grow gate oxide
4. Deposit poly
5. Etch poly

as can be seen from each line:

Step	Layer Name	Depth	Label	Comment
gd	-	2	-	# 3. Field oxide
e	Active	2	-	# 4.
gd	-	1	-	# 5. Gate oxide
gd	Poly	2	-	# 6. Polysilicon
e	NotPoly	2	-	# 7.

The active area drawn from this description is obtained by first growing a field oxide that is 2 units thick, and then selectively etching 2 units of oxide in regions specified by the layer Active. It should be mentioned that this type of isolation is different from standard LOCOS sequence described in Chapter 2 in that it does not yield a recessed oxide[1].

8.3 Sample File Description

The listing below describes the MOSIS Orbit 2-μm process (SCNA) that has been used throughout the book. This process has two poly layers and two metal layers. It may be found in the file morbn20.xst on the L-Edit disk.

```
# File: mORBn20.xst
# For: Cross-section process definition file
# Vendor: MOSIS:Orbit Semiconductor
# Technology: 2.0U N-Well (Lambda = 1.0um, Technology = SCNA)
```

---

[1] This is known as a "thick field oxide process."

```
# Technology Setup File: mORBn20.tdb
# Copyright (c) 1991-93
# Tanner Research, Inc. All rights reserved
# ************************************************************
# L-Edit
```

#Step	Layer Name	Depth	Label	Comment
gd		20	p-	#1. Substrate
id	"Well X"	8	n-	#2. n-Well
id	ActPSelNotPoly	2	p+	# 3. p-Implant
id	ActNSelNotPoly	2	n+	# 4. n-Implant
id	CCD&Act	1	-	# 5. CCD Implant
id	"P Base"	4	-	# 6. NPN Base Implant
gd	-	2	-	# 7. Field Oxide
e	Active	2	-	# 8.
gd	-	1	-	# 9. Gate Oxide
gd	Poly	2	-	# 10. Polysilicon
e	NotPoly	3	-	# 11.
gd	-	1	-	# 12. 2nd Gate Oxide
gd	Poly2	2	-	# 13. 2nd Polysilicon
e	NotPoly2	3	-	# 14.
gd	-	2	-	# 15.
e	"P/P2/Act Contact"	2	-	# 16.
gd	Metal1	2	-	# 17. Metal 1
e	"Not Metal1"	2	-	# 18.
gd	-	2	-	# 19.
e	Via	2	-	# 20.
gd	Metal2	2	-	# 21. Metal 2
e	"Not Metal2"	2	-	# 22.
gd	-	4	-	# 23. Overglass
e	Overglass	4	-	# 24.

The information provided in the file defines the layer order and characteristics that the cross-sectional viewer uses to construct the drawing.

The picture is drawn by L-Edit in the same order as the listing. This sequence can be observed when the cross-sectional viewer is activated. Alternately, each step can be viewed individually using the single-step option.

8.4 Cross-Sectional Viewer Files

The Student Version of L-Edit provides the following files for use with the cross-sectional viewer. These correspond to the technology setup files[1] that are provided

---

[1] See Chapter 2 for a description of the available technology files.

---

on disk.

MHP_N12.XST	The cross-section process file for MOSIS's Hewlett-Packard n-well 1.2 μm CMOS process. Technology = SCN, Lambda =0.6 μm.
MHP_N16.XST	The cross-section process file for MOSIS's Hewlett-Packard n-well 1.6 μm CMOS process. Technology = SCN, Lambda = 0.8 μm.
MORBN20.XST	The cross-section process file for MOSIS's Orbit Semiconductor n-well 2.0 μm CMOS process. Technology = SCNA, Lambda = 1.0 μm.
MORBP20.XST	The cross-section process file for MOSIS's Orbit Semiconductor p-well 2.0 μm CMOS process. Technology = SCPE, Lambda = 1.0 μm.
MVTIN20.XST	The cross-section process file for MOSIS's VLSI Technology n-well 2.0 μm CMOS process. Technology = SCN, Lambda = 1.0 μm.
ORBTN12.XST	The cross-section process file for Orbit Semiconductor's n-well 1.2 μm CMOS process. Technology = N122P2M, Rules = MOSIS_12.
ORBTN16.XST	The cross-section process file for Orbit Semiconductor's n-well 1.6 μm CMOS process. Technology = N162P2M, Rules = MOSIS_16.
ORBTN20.XST	The cross-section process file for Orbit Semiconductor's n-well 2.0 μm CMOS process. Technology = N202P2MNPN-BCCD, Rules = MOSIS_16.
ORBTP12.XST	The cross-section process file for Orbit Semiconductor's p-well 1.2 μm CMOS process. Technology = P122P2M, Rules = MOSIS_12.
ORBTP16.XST	The cross-section process file for Orbit Semiconductor's p-well 1.6 μm CMOS process. Technology = P162P2M, Rules = MOSIS_16.
ORBTP20.XST	The cross-section process file for Orbit Semiconductor's p-well 2.0 μm CMOS process. Technology = P202P2M, Rules = MOSIS_20.

It is important to ensure that the .xst file used to create the cross-sectional drawing is consistent with the technology base in L-Edit. The Cross-Sectional Viewer will give an error message if a mismatch exists.

8.5 Sample Viewing File

A sample file named xsect.tdb has been provided to illustrate the use of the cross-sectional viewer. This is a simple CMOS inverter with a process defined by the file xsect.xst. The view obtained for this file is shown in Figure 8.2. Note that intermediate oxide layers are shown explicitly in the drawing.

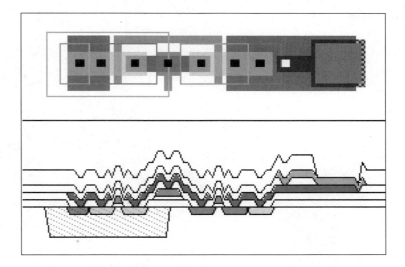

Figure 8.2. Example of cross-sectional viewer output.

Chapter 9

Layer Generation in L-Edit

Layer generation involves creating new layers from existing ones using various Boolean operations. This allows one to create single objects that consist of several layers.

9.1 Defining Derived Layers

Integrated circuits are constructed using a set of basis layers that are defined by the materials. A **derived layer** is constructed by altering one or more of the basis layers using a set of operations. L-Edit provides the operations AND, OR, NOT, GROW, and SHRINK for this purpose.

AND

The AND operation creates a new layer using two layers combined with the logic AND function in the form

Layer C= (Layer A) AND (Layer B).

The new layer consists of regions that are common to both Layer A and Layer B, as illustrated in the example of Figure 9.1. It is seen that the AND operation is simply

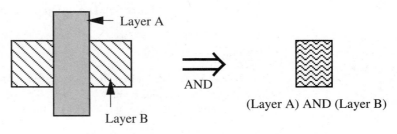

Figure 9.1. The AND operation with layers.

the intersection of the two objects.

OR

In a similar manner, the logical OR function creates a new layer using the expression

Layer D= (Layer A) OR (Layer B).

The resulting layer is the patterned regions from both Layer A and Layer B, merged into a single layer. An example of the OR operation is provided in Figure 9.2.

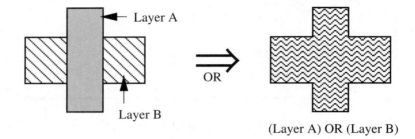

Figure 9.2. The OR operation with layers.

NOT

The NOT function creates a layer that has the complement of the original pattern. This is shown in Figure 9.3.

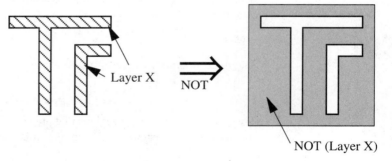

Figure 9.3. Definition of the NOT operation.

GROW

This operation creates a new layer from an existing one by increasing the size of all objects on the original layer by N units. The drawing in Figure 9.4 illustrates the effect of growing the size of a layer by 2 units.

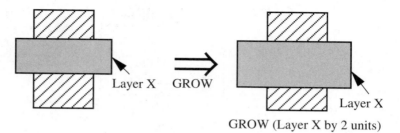

Figure 9.4. The GROW operation.

SHRINK

The SHRINK operation is the opposite of GROW. This option creates a new layer from an existing one by decreasing the size of all objects on the layer by a factor M units. An example is shown in Figure 9.5.

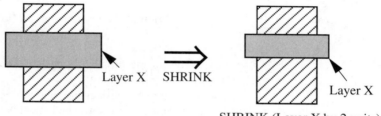

Figure 9.5. The SHRINK operation.

9.2 Examples from CMOS

The layout of devices in CMOS provides excellent examples of derived layers. Consider the SCNA technology that is loaded when L-Edit is launched. To create an n^+ region in the p-type substrate, we define

$$ndiff = (ACTIVE) \text{ AND } (NSELECT).$$

This can be understood by the physics of the fabrication discussed in Chapter 2. An ACTIVE area is one where the thin (gate) oxide is grown. NSELECT specifies the location of an n-type ion implant, but the implant can only penetrate the thin oxide. Thus, ndiff is an n^+ region that is created where ACTIVE and NSELECT overlap.

For a pMOSFET, we define p^+ regions using

$$pdiff = (ACTIVE) \text{ AND } (PSELECT) \text{ AND } (NWELL).$$

In this case, ACTIVE with PSELECT gives the region of the p^+ implant that makes it into the wafer. The additional specification of NWELL in the AND relationship

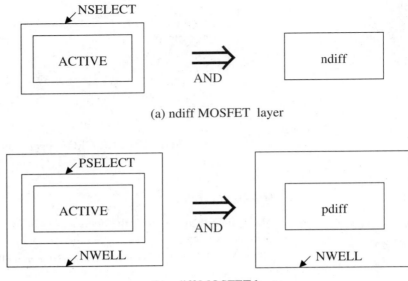

(a) ndiff MOSFET layer

(b) pdiff MOSFET layer

Figure 9.6. Formation of ndiff and pdiff layers.

places the pFET in the proper n-type background. The formation of n^+ and p^+ regions is illustrated in Figure 9.6.

As another example, consider the field regions of the chip where the thick field oxide is grown. This is defined by the logical operation

$$Field = NOT(ACTIVE).$$

Using this, we can see the obvious relation

$$chip\ surface = (ACTIVE) + (Field)$$

since this includes the entire surface.

A MOSFET can be defined in a similar manner. The gate of an nFET is described by

$$gate = (POLY)\ AND\ (ACTIVE).$$

An n-channel MOSFET can then be defined by

$$nTran = (gate)\ AND\ (NSELECT).$$

The drain and source regions are given by

$$SourceDrain = (ndiff)\ AND\ [NOT(gate)]$$

while the bulk (substrate) is obtained from

$$nTranBulk = NOT(NWELL).$$

The main regions of an nFET are shown in Figure 9.7. A p-channel MOSFET is similar, except that an NWELL must be included. This gives

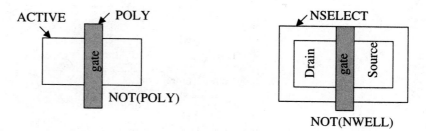

Figure 9.7. MOSFET formation using derived layers.

$$pTran= (gate) \text{ AND } (NWELL) \text{ AND } (PSELECT)$$

with

$$pTranBulk = NWELL$$

as the pFET bulk connection. A low-resistance well contact is constructed by creating an ndiff region in the n-well.

9.3 Layer Setup

To create a new layer, choose Layers from the Setup command group on the Menu Bar. This opens up the dialog window shown in Figure 9.8. The name of the layer

Setup

Palette...
Environment...

Layers...
Wires...
Special Layers...
Derived Layers...

Technology...
Frid...
Selection...

CIF...
GDS II...

DRC...

SPR Block
Padframes...
Pad Routes...

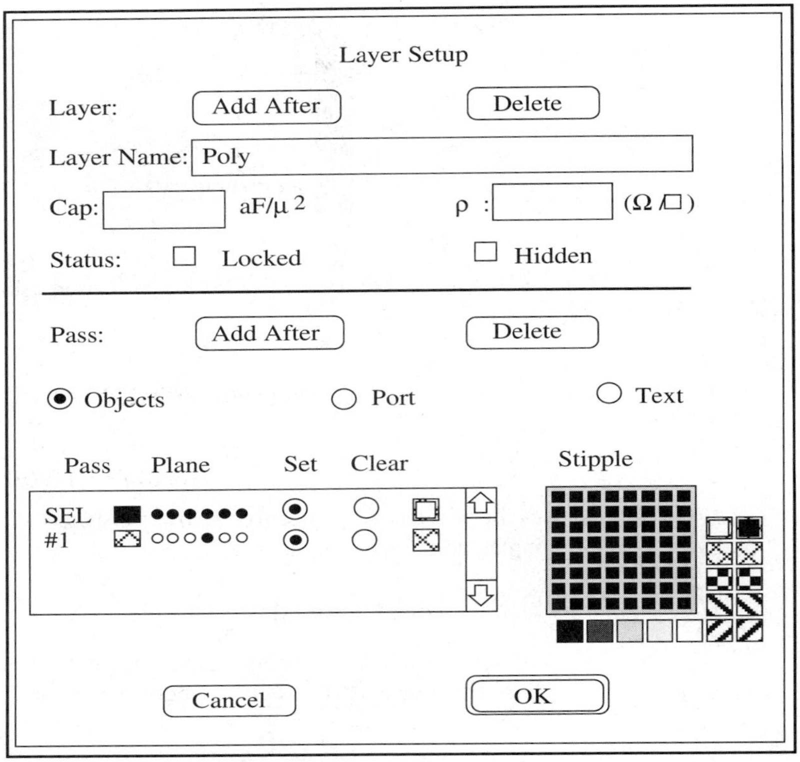

Figure 9.8. The Layers setup dialog window.

must be inserted into the layer list after the layers that are needed to derive it. For example, ndiff must occur after NSELECT and ACTIVE. The appearance of the layer on the L-Edit work area can be defined using the graphical editor tools in the lower right corner of the box. This dialog window also allows you to enter physical information, such as the capacitance and sheet resistance. Note in particular that the junction capacitance per unit area for the drain/source regions of an nMOSFET is entered on the ndiff layer to be used by the Extract algorithm. Similarly, pMOSFET depletion contributions must be entered on the pdiff layer.

After layer setup has been completed, choose Derived Layers from the Setup window. This opens the dialog box shown in Figure 9.9, allowing you to define the layer. Information on any layer can be accessed by pointing and clicking on the appropriate box in the Technology Palette. The example shows that the layer ndiff is obtained from the expression

$$ndiff = (ACTIVE) \text{ AND } (NSELECT),$$

as shown by the "AND" operation marked in the box. Arbitrary derived layers can be created in the same manner. A good exercise at this point is to point to layers in the Technology Palette and write the logical operations for them.

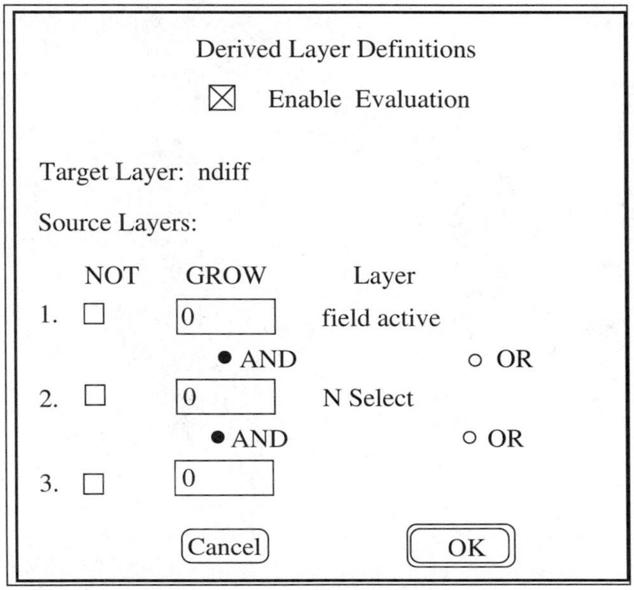

Figure 9.9. Layer definition dialog window.

9.4 Generate Layers Command

To invoke this feature, use the Special grouping on the Menu Bar. Choosing the command Generate Layers causes L-Edit to calculate all of the derived layers for the layout drawing. Layer Generation is automatically performed when you perform a DRC or Cross-Section View operation; consequently, the Generate Layers command is not normally used.

Chapter 10

Design Rule Checker

The L-Edit/DRC (design rule checker) is a program that examines the dimensions and spacings for every geometrical object in a layout, and looks for violations of the design rules. Design rule errors can be reported and identified directly on the layout, or a list can be written to a text file. The DRC allows you to check your work and make any necessary corrections before it is submitted for fabrication.

Design rule checking is accomplished by defining sets of rules for minimum sizes and spacings that apply to the fabrication process used in performing the layout.

10.1 Operation

The DRC routine is accessed from the Special grouping on the Menu Bar. This will start the design rule check on the current cell in the layout buffer. A dialog box will appear, prompting you to provide a name of the resulting DRC text file that will list all violations.

Special
Generate Layers...
Clear Gen'ed Layers...
DRC...
DRC Box...
Clear Error Layer...
Place and Route...
Extract...
Cross-Section...

Full-Chip DRC

The full-chip DRC option checks the entire layout for design rule violations. The DRC divides the layout into a grid of square *bins*, and then analyzes each bin for adherence to the layout rules. Separating the layout in this manner speeds up the analysis, since only local objects are compared.

Region-Only DRC

If you wish to run the DRC on a portion of the layout, use the DRC Box command. After this command is selected, the region to be checked is defined by drawing a box with the mouse. All objects that are contained in, or touch the boundary of, this box will be checked. This is particularly useful when a small section of a layout is modified.

Derived Layers

Both full-chip and region-only checks can be applied to derived layers created from the Generate Layer feature of L-Edit. For example, the drain and source n^+ regions in an n-channel MOSFET are defined by the derived layer

$$ndiff = (ACTIVE) \; AND \; (NSELECT)$$

which can be incorporated into the design rule set.

10.2 Error Reporting

Design rule violations are reported in three ways.

- Placing error ports at the location where a violation exists;
- Placing error markers at the location where a violation exists;
- Writing the violation to a text file.

Error ports act like standard L-Edit objects; they can be moved, examined, and deleted using the mouse. Error markers also have this property. Since the errors are shown on the layout, they can be identified. To remove the error layer, use the Clear Error Layer command. If a large number of design rule errors are found, the screen may get too cluttered to be useful. In this case, a text file output should be used.

If the design rule violations are reported in a text file, then the type of error and the coordinates are provided. Text file listings report errors using the format

<Rule Name>=<distance><unit name>; (<x1>,<y1>) -> (<x2>,<y2>)

where

<Rule Name> is the rule that was violated;
<distance> is the required value of the rule;
<unit name> specifies the units used in the rule;
(<x1>,<y1>) -> (<x2>,<y2> specifies the coordinates relative to the origin where the violation was found.

The text file listing will provide a separate line for each error, and will give the total number of violations found.

10.3 Design Rule Entry

Design rules are contained within the technology file that defines the current screen in the L-Edit buffer. The default design rule set in this version of L-Edit is SCNA, which loads from ledit.tdb. Any other technology file may be loaded using the **Replace Setup** command from the **File** window of the Menu Bar. This activates the dialog box shown in Figure 10.1. If you only want to load the DRC set, then click all switches OFF except for the DRC box.

Since the design rule set is automatically contained within the technology base, you may use an existing set by simply loading a layout file that has the appropriate information. If you change the name with the **Save As** command (under **File** in the Menu Bar), then the new file will have the same set.

L-Edit provides a direct interface for entering design rule values. This is accessed by choosing **DRC** from the **Setup** window in the Menu Bar and allows one to create an entire DR set as required. Executing this command brings the dialog box shown in Figure 10.2 to the center of the screen. Each layer may be selected, and numerical values entered using the keyboard. It is important to note that changes will be saved with the design file.

10.4 Design Rule Types

Geometrical spacings can be grouped into five basic categories: Minimum Width, Exact Width, Minimum Spacing, Surround, Overlap, and Extension. Each is discussed below.

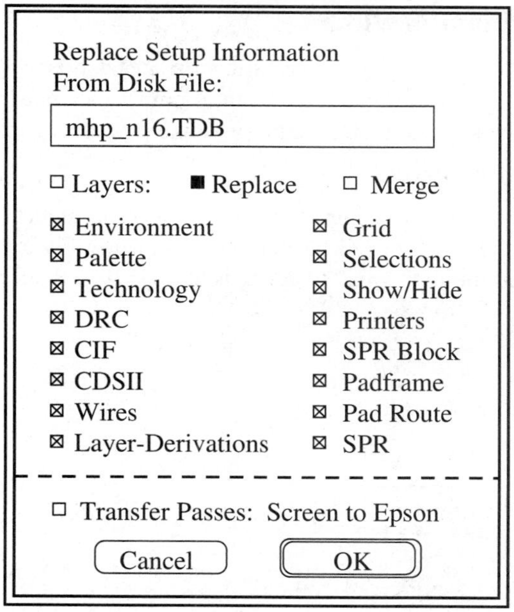

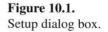

Replace Setup Information
From Disk File:

mhp_n16.TDB

□ Layers: ■ Replace   □ Merge

☒ Environment   ☒ Grid
☒ Palette   ☒ Selections
☒ Technology   ☒ Show/Hide
☒ DRC   ☒ Printers
☒ CIF   ☒ SPR Block
☒ CDSII   ☒ Padframe
☒ Wires   ☒ Pad Route
☒ Layer-Derivations   ☒ SPR

□ Transfer Passes: Screen to Epson

Cancel   OK

Figure 10.1.
Setup dialog box.

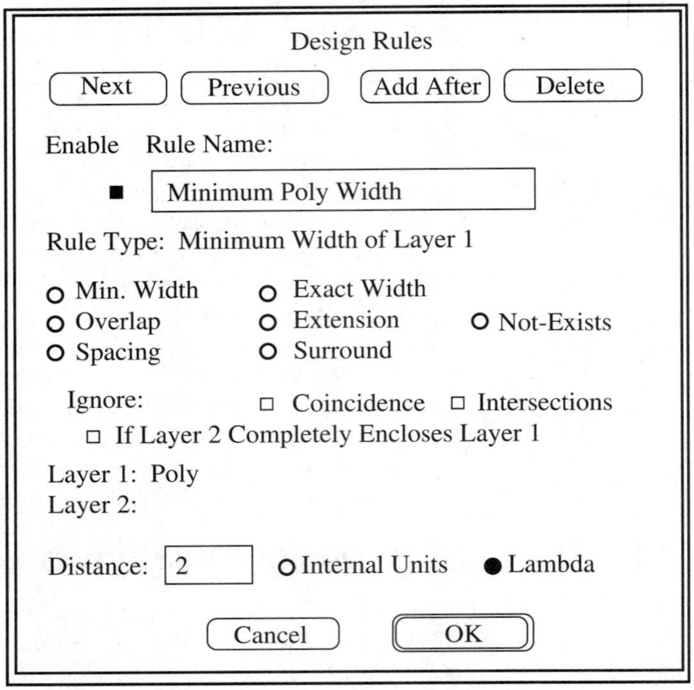

```
                    Design Rules
   ( Next )  ( Previous )   ( Add After )  ( Delete )

   Enable   Rule Name:

      ■    [ Minimum Poly Width              ]

   Rule Type:  Minimum Width of Layer 1

   O Min. Width      O Exact Width
   O Overlap         O Extension        O Not-Exists
   O Spacing         O Surround

      Ignore:          □ Coincidence  □ Intersections
         □ If Layer 2 Completely Encloses Layer 1

   Layer 1:  Poly
   Layer 2:

   Distance:  [ 2 ]    O Internal Units   ● Lambda

              ( Cancel )    (    OK    )
```

Figure 10.2. Dialog box for entering design rules.

Minimum Width

The minimum width specification says that an object must be at least W units wide as shown in Figure 10.3. A minimum width value applies to every layer, and is usually a limitation on the lithographic process.

Exact Width

This type of design rule says that certain objects are only permitted to have exact dimensions of (X × X) units, as shown in Figure 10.4. In a CMOS process, this is commonly applied to contacts and vias to enhance the yield of the process.

Figure 10.3. Minimum width specification.

Exact Width

Figure 10.4.
Exact width
specification.

Minimum Spacing

Minimum spacing rules give the smallest distance S between object edges on a layer, as illustrated in Figure 10.5. This type of rule is related to the lithographic resolution, and is also influenced by physical effects.

S   Minimum Spacing

Figure 10.5.
Minimum spacing
specification.

Surround

A surround rule applies when an object is embedded within another object. Figure 10.6 illustrates the minimum surround distance s that must exist between the square object and the larger rectangle. In a CMOS process, surround rules are important for contacts and vias. Physically, they allow for mask misalignment between different lithographic steps.

   s = Minimum Surround Spacing

Figure 10.6. Minimum surround specification.

Overlap and Extend

Overlap and extend rules are applied to two different layers. Both are illustrated in Figure 10.7. An extend rule gives the spacing E that an object must extend beyond the edge of another layer. An overlap rule is similar. It specifies the minimum distance O that objects on different layers must overlap. The most common example of this type of rule in CMOS is the gate overhang requirement of POLY with ndiff or pdiff to ensure proper formation of a MOSFET using the self-aligned gate process. In the SCNA technology, this is specified as an extension of POLY beyond ACTIVE/NSELECT.

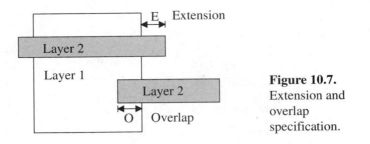

Figure 10.7. Extension and overlap specification.

10.5 Design Rules Listings

To obtain a listing of the design rules contained in the current file, execute the keyboard command

Alt-W

This creates an ASCII file named **filename.rul** on the disk drive. There is no equivalent command from the Menu Bar. The design rule listing is useful for understanding design rule violations that have been found by the DRC algorithm.

The text file generated by this command is made up of two major sections. The first section defines the layers that have been used with the format

LAYERNUMBER. (LAYER NAME)

The second section gives the actual values of the spacing rules and the layers involved. The general format for a DRC listing is as follows:

```
*********************************************
<Rule Name>
Type: <Rule Name>, Distance:<distance><unit name>
  Layer: <Layer Name>
  Layer: <Layer Name>
  Layer: <Layer Name>
*********************************************
```

where

- <Rule Name> is the name of the design rule
- <distance> is the value of the spacing/linewidth
- <unit> is the metric of the distance
- <Layer Name> gives the name of the layer(s) involved

The <Layer Name> listing is listed for every layer involved.

10.5.1 MOSIS 2-Micron Listing

The default technology in the Student Version of L-Edit is the MOSIS 2-micron scalable n-well (SCNA). The file that describes the process is listed below. The DRC description first defines the layers, and then lists each rule.

Layer Derivations and DRC Rules:
TDB file File0.TDB, Technology: MOSIS:Orbit 2U SCNA

Layer Setup:
1. (Grid Layer)
2. (Drag Box Layer)
3. (Origin Layer)
4. (Cell Outline Layer)
5. (Error Layer)
6. (Poly)
7. (Poly2)
8. (Active)
9. (Metal1)
10. (Metal2)
11. (N Well)
12. (N Select)
13. (P Select)
14. (Poly Contact)
15. (Poly2 Contact)
16. (ActiveContact)
17. (Via)
18. (CCD)
19. (P Base)
20. (Overglass)
21. (Pad Comment)
22. (Icon/Outline)
23. (Capacitor ID)
24. (Resistor ID)
25. (Label)
26. (Well X) = (N Well)
27. (Poly OrPoly2) = (Poly) OR (Poly2)
28. (NotPoly) = NOT(Poly)
29. (NotPoly2) = NOT(Poly2)
30. (ActNSelNotPoly) = (Active) AND (N Select) AND NOT(Poly OrPoly2)
31. (ActPSelNotPoly) = (Active) AND (P Select) AND NOT(Poly OrPoly2)
32. (P/P2/Act Contact) = (Poly Contact) OR (Poly2 Contact) OR (ActiveContact)
33. (Not Metal1) = NOT(Metal1)
34. (Not Metal2) = NOT(Metal2)
35. (CCD&Act) = (CCD) AND (Active)
36. (poly wire) = (Poly) AND NOT(Resistor ID)

37. (poly res) = (Poly) AND (Resistor ID)
38. (subs) = NOT(N Well)
39. (NotPBase) = NOT(P Base)
40. (all poly) = (Poly) OR (Poly2)
41. (gate1) = (Poly) AND (Active)
42. (gate2) = (Poly2) AND NOT(Poly) AND (Active)
43. (field active) = NOT(all poly) AND (Active)
44. (pdiff) = (field active) AND (P Select)
45. (ndiff) = (field active) AND (N Select)
46. (ntran) = (gate1) AND (subs)
47. (ptran) = (gate1) AND (N Well)
48. (ntran2) = (gate2) AND (subs)
49. (ptran2) = (gate2) AND (N Well)
50. (Gates) = (gate1) OR (gate2)
51. (Field Poly) = (Poly) AND NOT(Active)
52. (n Active) = (Active) AND (N Select)
53. (p Active) = (Active) AND (P Select)
54. (SelNotInWell) = (N Select) AND NOT(N Well)
55. (OtherSelInWell) = (P Select) AND (N Well)
56. (pSelect&nSelect) = (P Select) AND (N Select)
57. (Not Selected Active) = (Active) AND NOT(N Select) AND NOT(P Select)
58. (ViaNotOnPad) = (Via) AND NOT(Pad Comment)
59. (n Active InPBase) = (P Base) AND (n Active)
60. (allsubs) = (subs) OR (N Well)
61. (P1&P2 & Poly2Cap) = (Capacitor ID) AND (Poly) AND (Poly2)

Design Rules:

```
****************************************************
** DISABLED **
* MOSIS:ORBIT 2.0U SCNA *
Type: Minimum Width, Distance: -1 Lambda
****************************************************
```

1.1 Well Minimum Width
   Type: Minimum Width, Distance: 10 Lambda
      Layer: N Well
```
****************************************************
```

1.3 Well to Well(Same Potential) Spacing
   Type: Spacing, Distance: 6 Lambda
      Layer: N Well
```
****************************************************
```

2.1 Active Minimum Width
   Type: Minimum Width, Distance: 3 Lambda
      Layer: Active
```
****************************************************
```

2.2 Active to Active Spacing
Type: Spacing:I:E, Distance: 3 Lambda
  Layer: Active
****

2.3a Source/Drain Active to Well Edge
Type: Surround:O, Distance: 5 Lambda
  Layer: pdiff
    Layer: field active
      NOT Layer: all poly
        Layer: Poly
        OR
        Layer: Poly2
      AND
      Layer: Active
    AND
    Layer: P Select
  Layer: N Well
****

2.3b Source/Drain Active to Well Space
Type: Spacing:E, Distance: 5 Lambda
  Layer: ndiff
    Layer: field active
      NOT Layer: all poly
        Layer: Poly
        OR
        Layer: Poly2
      AND
      Layer: Active
    AND
    Layer: N Select
  Layer: N Well
****

2.4a WellContact(Active) to Well Edge
Type: Surround:O, Distance: 3 Lambda
  Layer: n Active
    Layer: Active
    AND
    Layer: N Select
  Layer: N Well
****

2.4b SubsContact(Active) to Well Spacing
Type: Spacing:E, Distance: 3 Lambda
  Layer: p Active
    Layer: Active
    AND
    Layer: P Select

Layer: N Well
```
*****************************************************
```

3.1 Poly Minimum Width
Type: Minimum Width, Distance: 2 Lambda
Layer: Poly
```
*****************************************************
```

3.2 Poly to Poly Spacing
Type: Spacing:I:E, Distance: 2 Lambda
Layer: Poly
```
*****************************************************
```

3.3 Gate Extension out of Active
Type: Extension, Distance: 2 Lambda
Layer: Active
Layer: Poly
```
*****************************************************
```

3.4a/4.1a Source/Drain Width
Type: Extension, Distance: 3 Lambda
Layer: all poly
    Layer: Poly
    OR
    Layer: Poly2
Layer: n Active
    Layer: Active
    AND
    Layer: N Select
```
*****************************************************
```

3.4b/4.1b Source/Drain Width
Type: Extension, Distance: 3 Lambda
Layer: all poly
    Layer: Poly
    OR
    Layer: Poly2
Layer: p Active
    Layer: Active
    AND
    Layer: P Select
```
*****************************************************
```

3.5 Poly to Active Spacing
Type: Spacing:I:E, Distance: 1 Lambda
Layer: Poly
Layer: Active
```
*****************************************************
```

4.2a Active(In Select) to Select Edge
Type: Surround:I:O, Distance: 2 Lambda
Layer: Active
Layer: SelNotInWell

Layer: N Select
AND
NOT Layer: N Well
****

4.2b Active(InOtherSel) to OtherSel Edge
Type: Extension, Distance: 2 Lambda
Layer: Active
Layer: OtherSelInWell
Layer: P Select
AND
Layer: N Well
****

4.3a Select Edge to ActCnt
Type: Surround:O, Distance: 1 Lambda
Layer: ActiveContact
Layer: ndiff
Layer: field active
NOT Layer: all poly
Layer: Poly
OR
Layer: Poly2
AND
Layer: Active
AND
Layer: N Select
****

4.3b Select Edge to ActCnt
Type: Surround:O, Distance: 1 Lambda
Layer: ActiveContact
Layer: pdiff
Layer: field active
NOT Layer: all poly
Layer: Poly
OR
Layer: Poly2
AND
Layer: Active
AND
Layer: P Select
****

4.4a Select Minimum Width
Type: Minimum Width, Distance: 2 Lambda
Layer: N Select
****

4.4b Select Minimum Width
Type: Minimum Width, Distance: 2 Lambda

Layer: P Select
```
****************************************************
```

4.4c Select to Select Spacing

Type: Spacing, Distance: 2 Lambda
Layer: N Select
```
****************************************************
```

4.4d Select to Select Spacing

Type: Spacing, Distance: 2 Lambda
Layer: P Select
```
****************************************************
```

4.5 Not Existing: pSel Overlap of nSel

Type: NotExists, Distance: 0 Lambda
Layer: pSelect&nSelect
Layer: P Select
AND
Layer: N Select
```
****************************************************
```

4.6 Not Existing: Not Selected Active

Type: NotExists, Distance: 2 Lambda
Layer: Not Selected Active
Layer: Active
AND
NOT Layer: N Select
AND
NOT Layer: P Select
```
****************************************************
```

5A.1 Poly Contact Exact Size

Type: Exact Width, Distance: 2 Lambda
Layer: Poly Contact
```
****************************************************
```

5A.2/5B.6 FieldPoly Overlap of PolyCnt

Type: Surround, Distance: 2 Lambda
Layer: Poly Contact
Layer: Field Poly
Layer: Poly
AND
NOT Layer: Active
```
****************************************************
```

5A.3 PolyContact to PolyContact Spacing

Type: Spacing, Distance: 2 Lambda
Layer: Poly Contact
```
****************************************************
```

6A.1 Active Contact Exact Size

Type: Exact Width, Distance: 2 Lambda
Layer: ActiveContact
```
****************************************************
```

6A.2/6A.4 FieldActive Overlap of ActCnt

Type: Surround, Distance: 2 Lambda
    Layer: ActiveContact
    Layer: field active
      NOT Layer: all poly
        Layer: Poly
        OR
        Layer: Poly2
      AND
      Layer: Active

****

6A.3 ActCnt to ActCnt Spacing

Type: Spacing, Distance: 2 Lambda
    Layer: ActiveContact

****

7.1 Metal1 Minimum Width

Type: Minimum Width, Distance: 3 Lambda
    Layer: Metal1

****

7.2 Metal1 to Metal1 Spacing

Type: Spacing, Distance: 3 Lambda
    Layer: Metal1

****

7.3 Metal1 Overlap of PolyContact

Type: Surround, Distance: 1 Lambda
    Layer: Poly Contact
    Layer: Metal1

****

7.4 Metal1 Overlap of ActiveContact

Type: Surround, Distance: 1 Lambda
    Layer: ActiveContact
    Layer: Metal1

****

8.1 Via Exact Size

Type: Exact Width, Distance: 2 Lambda
    Layer: ViaNotOnPad
      Layer: Via
      AND
      NOT Layer: Pad Comment

****

8.2 Via to Via Spacing

Type: Spacing, Distance: 3 Lambda
    Layer: Via

****

8.3 Metal1 Overlap of Via

Type: Surround, Distance: 1 Lambda

Layer: Via
Layer: Metal1
****

8.4a Via to Poly Spacing

Type: Spacing:E, Distance: 2 Lambda
Layer: ViaNotOnPad
Layer: Via
AND
NOT Layer: Pad Comment
Layer: Poly
****

8.4b Via(On Poly) to Poly Edge

Type: Surround:O, Distance: 2 Lambda
Layer: Via
Layer: Poly
****

8.4c Via to Active Spacing

Type: Spacing:E, Distance: 2 Lambda
Layer: Via
Layer: Active
****

8.4d Via (On Active) to Active Edge

Type: Surround:O, Distance: 2 Lambda
Layer: Via
Layer: Active
****

8.5a Via to PolyContact Spacing

Type: Spacing, Distance: 2 Lambda
Layer: Via
Layer: Poly Contact
****

8.5b Via to ActiveContact Spacing

Type: Spacing, Distance: 2 Lambda
Layer: Via
Layer: ActiveContact
****

9.1 Metal2 Minimum Width

Type: Minimum Width, Distance: 3 Lambda
Layer: Metal2
****

9.2 Metal2 to Metal2 Spacing

Type: Spacing, Distance: 4 Lambda
Layer: Metal2
****

9.3 Metal2 Overlap of Via

Type: Surround, Distance: 1 Lambda

Layer: Via
Layer: Metal2
****

** DISABLED **

10.1a Bonding Area: OverGlass (88x88um)

Type: Exact Width, Distance: 88 Lambda
Layer: Overglass
****

10.1b BondingArea:PadComment(100x100um)

Type: Surround, Distance: 6 Lambda
Layer: Overglass
Layer: Pad Comment
****

10.1c Bonding Area: Via (90x90um)

Type: Surround:O, Distance: 1 Lambda
Layer: Overglass
Layer: Via
****

10.1d Bonding Area: Metal2 (100x100um)

Type: Surround, Distance: 6 Lambda
Layer: Overglass
Layer: Metal2
****

10.1e Bonding Area: Metal1 (102x102um)

Type: Surround, Distance: 7 Lambda
Layer: Overglass
Layer: Metal1
****

** DISABLED **

10.4 Pad to Unrelated-Metal2 Space(30um)

Type: Spacing:I:E, Distance: 30 Lambda
Layer: Pad Comment
Layer: Metal2
****

** DISABLED **

10.5a Pad to Unrelated-Metal1 (15um)

Type: Spacing:I:E, Distance: 15 Lambda
Layer: Pad Comment
Layer: Metal1
****

** DISABLED **

10.5b Pad to Unrelated-Poly Space (15um)

Type: Spacing, Distance: 15 Lambda
Layer: Pad Comment
Layer: Poly
****

** DISABLED **
10.5c Pad to Unrelated-Act Space (15um)
 Type: Spacing, Distance: 15 Lambda
  Layer: Pad Comment
  Layer: Active
****

11.1 Capacitor: Poly2 Minimum Width
 Type: Minimum Width, Distance: 3 Lambda
  Layer: P1&P2 & Poly2Cap
   Layer: Capacitor ID
   AND
   Layer: Poly
   AND
   Layer: Poly2
****

11.2/12.2 Cap/Trans: Poly2toPoly2 Space
 Type: Spacing, Distance: 3 Lambda
  Layer: Poly2
****

11.4a Capacitor: Poly2 Space to Active
 Type: Spacing, Distance: 2 Lambda
  Layer: P1&P2 & Poly2Cap
   Layer: Capacitor ID
   AND
   Layer: Poly
   AND
   Layer: Poly2
  Layer: Active
****

11.4b Capacitor: Poly2 to Well Spacing
 Type: Spacing:E, Distance: 2 Lambda
  Layer: P1&P2 & Poly2Cap
   Layer: Capacitor ID
   AND
   Layer: Poly
   AND
   Layer: Poly2
  Layer: N Well
****

11.4c Capacitor: Poly2 to Well Edge
 Type: Surround:O, Distance: 2 Internal Units
  Layer: P1&P2 & Poly2Cap
   Layer: Capacitor ID
   AND
   Layer: Poly
   AND

Layer: Poly2
Layer: N Well
****

11.5/12.6a Cap/T: Poly2 Space to PolyCnt
Type: Spacing:E, Distance: 3 Lambda
Layer: Poly Contact
Layer: Poly2
****

12.1 Transistor: Poly2 Minimum Width
Type: Minimum Width, Distance: 2 Lambda
Layer: Poly2
****

12.3 Trans: Gate Extension Out of Act
Type: Extension, Distance: 2 Lambda
Layer: Active
Layer: Poly2
****

12.4 Transistor: Poly2 to Active Spacing
Type: Spacing:I:E, Distance: 1 Lambda
Layer: Poly2
Layer: Active
****

12.5a Transistor: Poly2 to Poly Spacing
Type: Spacing:I:E, Distance: 2 Lambda
Layer: Poly2
Layer: Poly
****

12.5b Cap/Trans: Poly Overlap of Poly2
Type: Overlap, Distance: 2 Lambda
Layer: Poly2
Layer: Poly
****

12.5c Trans: P1&P2overlap to P2Edge
Type: Extension, Distance: 2 Lambda
Layer: Poly
Layer: Poly2
****

11.3/12.5d Cap/T: P1&P2overlap to P1Edge
Type: Extension, Distance: 2 Lambda
Layer: Poly2
Layer: Poly
****

12.6b Transistor: Poly2 to ActCnt Space
Type: Spacing, Distance: 3 Lambda
Layer: ActiveContact
Layer: Poly2

```
**************************************************
```

13.1 Poly2Contact Exact Size
Type: Exact Width, Distance: 2 Lambda
  Layer: Poly2 Contact

```
**************************************************
```

13.2 Poly2Contact to Poly2Cnt Spacing
Type: Spacing, Distance: 2 Lambda
  Layer: Poly2 Contact

```
**************************************************
```

13.3 P2CapPlates Overlap of Poly2Cnt
Type: Surround:O, Distance: 3 Lambda
  Layer: Poly2 Contact
  Layer: P1&P2 & Poly2Cap
    Layer: Capacitor ID
    AND
    Layer: Poly
    AND
    Layer: Poly2

```
**************************************************
```

13.4 Poly2(NotCapPlate) Overlap Poly2Cnt
Type: Surround, Distance: 2 Lambda
  Layer: Poly2 Contact
  Layer: Poly2

```
**************************************************
```

13.5a Poly2Cnt Space to Poly
Type: Spacing:E, Distance: 3 Lambda
  Layer: Poly2 Contact
  Layer: Poly

```
**************************************************
```

13.5b Poly2Cnt Space to Active
Type: Spacing, Distance: 3 Lambda
  Layer: Poly2 Contact
  Layer: Active

```
**************************************************
```

13.5c Poly2Cnt Space to Via
Type: Spacing, Distance: 2 Lambda
  Layer: Poly2 Contact
  Layer: Via

```
**************************************************
```

13.6 Metal1 Overlap Poly2Cnt
Type: Surround, Distance: 1 Lambda
  Layer: Poly2 Contact
  Layer: Metal1

```
**************************************************
```

14.1 N-Active (Emittor) Surround ActCnt
Type: Surround:O, Distance: 3 Lambda

            Layer: ActiveContact
            Layer: n Active InPBase
               Layer: P Base
               AND
               Layer: n Active
               Layer: Active
                  AND
                  Layer: N Select
****

14.2 P-Base surround N-Active (Emittor)
   Type: Surround, Distance: 2 Lambda
         Layer: n Active InPBase
            Layer: P Base
            AND
            Layer: n Active
               Layer: Active
               AND
               Layer: N Select
         Layer: P Base
****

14.3 N-Active (Emittor) to PBaseCnt
   Type: Spacing:E, Distance: 4 Lambda
         Layer: p Active
            Layer: Active
            AND
            Layer: P Select
         Layer: n Active InPBase
            Layer: P Base
            AND
            Layer: n Active
               Layer: Active
               AND
               Layer: N Select
****

14.4 P-Base to Eage of N-Well
   Type: Surround:O, Distance: 6 Lambda
      Layer: P Base
      Layer: N Well
****

14.5 P-Base space to N-Active (WellCnt)
   Type: Spacing:E, Distance: 4 Lambda
         Layer: n Active
            Layer: Active
            AND
            Layer: N Select
         Layer: P Base

```
*****************************************************
```

14.6 N-Select surround N-Active(Emitter)
  Type: Surround, Distance: 2 Lambda
    Layer: n Active InPBase
      Layer: P Base
      AND
      Layer: n Active
        Layer: Active
        AND
        Layer: N Select
    Layer: N Select

```
*****************************************************
```

Chapter 11

Circuit Extraction Using L-Edit

Circuit extraction is a powerful feature of L-Edit that translates a layout drawing into a SPICE-compatible text file for circuit simulation using any standard version of SPICE. This chapter provides the details needed to understand and use the Extract feature of the L-Edit program. Most of the chapter was taken directly from the L-Edit Manual to ensure completeness and accuracy.

11.1 Operation

The process of circuit extraction relies on defining groups of geometrical objects that physically represent distinct electronic elements. The program searches the layout to locate and identify the pattern groups, and then acts as a translator by creating an element description line that characterizes the section. Information on the type of element, its connections, and important parameters are included for each element.

To access the circuit extractor, use the Special grouping from the Menu bar and select Extract. A dialog box will appear, and you will be prompted for the name of the data file that contains the extraction information. These are identified by filenames with a .ext extension. For example, transistors in the SCNA technology are described by the file named **morbn20.ext**. You must also type in the filename that you want to use for the SPICE listing. In L-Edit, these have a default extension of **.spc**.

Once you have provided both the extraction file name and the output (SPICE) file name, click "OK" in the dialog window. L-Edit will analyze the layout drawing, determine the location and connections of all devices that are defined in the .ext file, and print a SPICE-compatible ASCII output listing in the .spc file. L-Edit will assign names to every device (such as M2 for a MOSFET) and provide node num-

bers as needed.

The resulting SPICE-compatible file is made up of (a) element statements corresponding to the defined devices that have been identified, (b) .MODEL lines for every active device, and (c) comment lines. The .MODEL lines do not contain any data, but are printed to remind you to add the parameters for each device. Also, it is important to note that L-Edit cannot identify the ground node from your layout. Before using the listing in a SPICE simulation, it is necessary to identify the ground node as required in a node analysis. This can be done explicitly by finding and changing every occurrence, or implicitly by adding a resistor with a value of $R=0$ ohms between the numbered node and ground (0).

11.2 Description of the Extraction Process

Layout extraction algorithms are designed to locate and identify predefined groups of objects that correspond to physical implementations of lumped-element electronic devices. In addition, interconnect patterns are translated into wires to complete the circuit description. The resulting file, called a **netlist,** provides a listing of every element and how it is connected in the circuit.

L-Edit/Extract is a process-independent layout extractor capable of recognizing a variety of passive and active devices. The algorithm is integrated into the L-Edit environment, and is capable of extracting a full chip. It utilizes L-Edit's Generate Layers feature to perform logical operations on layers, and it outputs its results in standard SPICE (Berkeley 2G6) netlist format.

11.2.1 Device Extraction

Once a layout has been created, an extractor can be used to determine the circuit represented by the layout. This circuit can then be used by other tools to verify or simulate the layout. L-Edit/Extract is a generic device extractor. It is capable of recognizing active devices including BJTs, diodes, GaAsFETs, JFETs and MOSFETs. It can also extract passive devices: capacitors, inductors, and resistors. The extractor is also capable of extracting non-standard devices through the SPICE subcircuit construction. In addition, L-Edit/Extract is process-independent. This is accomplished by defining a series of connections that describe how the layers interact electrically.

11.2.2 Full-Chip Extraction

L-Edit/Extract is capable of extracting large regions of layout. The extractor divides the layout into a grid of square "bins" and extracts each bin. This process significantly increases the performance, because of reduced memory requirements to process each bin; performance decreases rapidly as the number of objects to be considered simultaneously increases. The bin size is set through the Extract dialog box, and is given in internal units.

11.2.3 Derived Layers

The extractor definitions used by L-Edit/Extract can be specified using any derived layer from L-Edit's Generate Layers feature. This capability greatly expands the set of rules that can be specified. Intermediate layers can be temporarily generated using AND, OR, and NOT operations, and the derived results can be utilized in a design rule. The derived layers are generated and disposed of automatically. Please note that L-Edit/Extract does not currently utilize any GROW parameters associated with derived layers.

11.2.4 Parameter Extraction

L-Edit/Extract is capable of extracting the most common device parameters, including resistance, capacitance, and device length, width, and area. These parameters provide useful information when verifying drive, fanout, and other circuit performance characteristics.

11.2.5 SPICE Output

L-Edit/Extract creates a netlist file that contains the circuit description. This netlist is in Berkeley 2G6 SPICE format, so that it can be used with a SPICE simulator, Tanner Tools' LVS, or other tools that read SPICE format netlists. The Extractor's run time is also printed out in this file.

11.3 Why Extract?

Design rule checking ensures that a layout conforms to fabrication process requirements, but it does not verify that the layout actually implements what was intended, nor does it assist in determining whether the circuit will perform to specification. These problems require the technique of layout extraction. Layout extraction is the act of analyzing a piece of layout and constructing the circuit it represents. Extraction produces a netlist that contains a representation of the devices and connectivity of the layout. This netlist can be compared against another netlist in a process called netlist comparison. If the extracted netlist is compared against one generated from a schematic, the netlist comparison process is also

---

referred to as layout versus schematic, or LVS. LVS ensures that the layout circuit is equivalent to the schematic circuit.

In addition to device and connectivity information, the extracted netlist contains parametric information about the devices constructed in the layout, so that it is also possible to use the netlist for device-level simulations. The simulation aids in verifying device sizes, drive capabilities, and running other circuit performance tests.

11.4 Configuring the Extractor

There are three steps involved in configuring the extractor: deciding on the set of devices and connections that are to be extracted, setting up the layers which uniquely define the desired devices and connections, and entering the devices and connections into the extractor definition file.

11.4.1 Devices and Connections

The first step in configuring the extractor is to decide on the set of devices and connections that need to be extracted. A device is any circuit element, such as a transistor, resistor, capacitor, or diode. A connection is any electrical connectivity between two process layers, such as between Poly and Metal1 when a contact is present.

You only need to define those devices and connections that are of interest to you. For instance, every design contains resistors because no process layer is a perfect conductor. However, you might know that your circuit does not contain any wire long enough for its inherent resistance to impact the circuit's performance; hence, you would not need to extract wires as resistor devices. Or you might be using a process that has two metal layers, but your design actually uses only one of these layers. You would then not need to extract the connection between the first and second metal layer. Once you have decided on the set of devices and connections that need to be extracted, you can then move on to the next step.

11.4.2 Setting Up Layers

To extract devices, you must instruct the extractor how to recognize and interpret the devices. L-Edit itself has no knowledge of semiconductors or other devices. It simply assists you in arranging objects from different layers into a pattern. You must then teach L-Edit/Extract the meaning of the pattern that you have constructed. This is accomplished by using L-Edit's Generate Layers feature to construct a set of layers that uniquely define the devices and their various pins. We will now illustrate a specific example. This example defines everything that is needed to extract transistors from a CMOS N Well process. You would start by constructing a recognition layer for each device. A recognition layer is a layer that uniquely defines the existence of a device. For instance, you would define the gate of a MOS transistor as the intersection of a Poly region with an Active region. However, the gate is not sufficient to uniquely identify a transistor, because it does not discriminate between N- and P-type MOS devices. To further distinguish between these types of transistors, you must specify what type of well contains the transistor. This is done by defining an ntran layer and a ptran layer. Here is how you would implement this example:

Given the following process layers

Poly, Active, N Well, N Select, P Select,

we define the MOSFET regions as follows:

Gate	= (Poly) AND (Active)
Substrate	= NOT(N Select) AND (N Well)
N Well	= NOT(N Select) AND (N Well)
N Select	= NOT(N Well) AND (N Select)
P Select	= NOT(N Well) AND (P Select)
nTran	= (Gate) AND (Select AND (NOT(N Well)))
pTran	= (Gate) AND (Well) AND (NOT(N Select))
ndiff	= (Field Active) AND (N Select)
pdiff	= (Field Active) AND (P Select)

Once you have defined the recognition layer, you can then define the remaining layers needed for the device. The required layers are described in the Using Extractor Definitions section.

For resistors and capacitors, you will also need to supply the extractor with a capacitance per unit area (in units of attoFarads (10^{-18}) per square micron), and a resistance value (in units of ohms per unit area) for each layer. These two numbers are entered into the Setup Layers dialog chosen from the Setup Menu. Capacitance is computed as the area of the recognition layer multiplied by the capacitance constant for the given layer. Resistance is computed by multiplying the resistance constant by the length of the resistor, then dividing by the width. Note that it is important to use the WIDTH parameter in the extractor definition file to correctly compute resistance values.

11.4.3 Extractor Definition File

Once you have defined all of the layers needed to identify the devices, you can enter the device information into the extractor definition file. The extractor definition file contains a list of the connections and devices that are to be extracted. The example below lists the contents of the example.ext extractor definition file that is shipped on the L-Edit disk.

```
connect (Poly, Metal1, PolyContact)
connect (ndiff, Metal1, ActiveContact)
connect (Metal1, Metal2, Via)
connect (pdiff, Metal1, ActiveContact)

# Substrate contact connect (Substrate, pdiff, pdiff)
# Well contact connect (N Well, ndiff, ndiff)

device = MOSFET(
```

```
                    RLAYER=nTran;
                    Drain=ndiff, WIDTH;
                    Gate=Poly;
                    Source=ndiff, WIDTH;
                    Bulk=Substrate;
                    MODEL=NMOS;
                    )


    device = MOSFET(

                    RLAYER=pTran;

                    Drain=pdiff, WIDTH;

                    Gate=Poly;

                    Source=pdiff, WIDTH;

                    Bulk=N Well;

                    MODEL=PMOS;

                    )
```

11.5 Running the Extractor

The extractor is invoked by choosing Extract from the Special menu. Enter the name of the extractor definition file and the name of the SPICE output file. Define the bin size the Extractor will use to decompose the layout to be extracted. Select Write Node Names and Write Node Capacitances if desired. Note that L-Edit/ Extract only operates on boxes and orthogonal polygons and orthogonal wires, and thus cannot extract circles.

Enter the names of the extractor definition file and the SPICE Extractor output file that will be used to write the results of the extraction. In order to increase Extractor performance, large regions of layout are divided into smaller sections, which are then handled individually. These components are called bins, and are square in shape. The bin size is given in internal units, and defaults to 100 units by 100 units.

If Write Node Names is selected, then L-Edit/Extract will examine port objects on the layout and write the port names into the output file as SPICE comments. In order for a port to be written into the output file, it must be completely contained by a node polygon, and it must be on the same layer as the node polygon.

If Write Node Capacitances is selected, then L-Edit/Extract will compute the capacitance of each node in the circuit using the capacitance numbers specified in the Setup Layers dialog box under the Setup menu. The capacitance is included in the output netlist by including an appropriate-valued capacitor for each node in the circuit: one pin of the capacitor is attached to the corresponding node, and the other pin is attached to a node named '0' (SPICE ground name). L-Edit/Extract is not capable of determining whether other nodes in the circuit are ground nodes, so this '0' node will need to be manually connected to other ground nodes. This can be accomplished by editing the SPICE output file (find all ground nodes and rename them to be '0'); some SPICE simulators are also capable of defining two nodes as equivalent (refer to your SPICE reference manual). After entering the above information into the dialog, click OK. L-Edit/Extract will then extract the current cell, including any instances contained in the cell. A standard SPICE netlist will be produced as output. It will contain all of the extracted nodes and devices in SPICE format. In addition, an informational comment will follow each device. Here is an example:

C1 1266 1269 259.6FF

*Pos Neg (10 15 19 22) A=63

The first line is the SPICE statement referencing a 259.6FF capacitor with the unique name of 1, which has one pin connected to node 1266 and the other pin connected to node 1269. The next line begins with the SPICE comment character (*) and then contains the label Pos Neg. The order of this list is the same order as in the line above, so that in this example the Pos pin is connected to node 1266 and the Neg pin to node 1269. Following this pin list are four numbers contained in a set of parentheses. These numbers correspond to the left, bottom, right, and top boundaries, respectively, of the device's recognition layer. For this example, the left-most edge of the capacitor's recognition layer is at X=10, and the right-most edge at X=19. Note that these four numbers are not coordinate pairs, because the recognition layer may not be rectangular.

11.6 Student L-Edit Extraction Files

This version of L-Edit provides several files for use with the corresponding technology base options.

MHP_N12.EXT   The extractor definition file for MOSIS's Hewlett-Packard n-well 1.2 μm CMOS process. Technology = SCN, Lambda =0.6 μm.

MHP_N16.EXT   The extractor definition file for MOSIS's Hewlett-Packard n-well 1.6 μm CMOS process. Technology = SCN, Lambda = 0.8 μm.

MORBN20.EXT   The extractor definition file for MOSIS's Orbit Semiconductor n-well 2.0 μm CMOS process. Technology = SCNA, Lambda = 1.0 μm.

MORBP20.EXT   The extractor definition file for MOSIS's Orbit Semiconductor p-well 2.0 μm CMOS process. Technology = SCPE, Lambda = 1.0 μm.

MVTIN20.EXT   The extractor definition file for MOSIS's VLSI Technology n-well 2.0 μm CMOS process. Technology = SCN, Lambda = 1.0 μm.

ORBTN12.EXT   The extractor definition file for Orbit Semiconductor's n-well 1.2 μm CMOS process. Technology = N122P2M, Rules = MOSIS_12.

ORBTN16.EXT   The extractor definition file for Orbit Semiconductor's n-well 1.6 μm CMOS process. Technology = N162P2M, Rules = MOSIS_16.

ORBTN20.EXT   The extractor definition file for Orbit Semiconductor's n-well 2.0 μm CMOS process. Technology = N202P2MNPNB-CCD, Rules = MOSIS_16.

ORBTP12.EXT   The extractor definition file for Orbit Semiconductor's p-well 1.2 μm CMOS process. Technology = P122P2M, Rules = MOSIS_12.

ORBTP16.EXT   The extractor definition file for Orbit Semiconductor's p-well 1.6 μm CMOS process. Technology = P162P2M, Rules = MOSIS_16.

ORBTP20.EXT   The extractor definition file for Orbit Semiconductor's p-well 2.0 μm CMOS process. Technology = P202P2M, Rules = MOSIS_20.

The file used in the extraction process must be the same as the technology used to create the layout.

The listing of the morbn20.ext file is shown below. Note that the following devices are defined

- nMOS transistor (POLY1 gate)
- pMOS transistor (POLY2 gate)
- nMOS transistor (POLY2 gate)
- pMOS transistor (POLY2 gate)
- NPN bipolar transistor
- POLY resistor
- POLY-to-POLY capacitor

using the associated statements defining connections among the layers.

```
# File: Morbn20.ext
# For: Extractor definition file
# Vendor: MOSIS:Orbit Semiconductor
# Technology: 2.0U N-Well (Lambda = 1.0um, Technology = SCNA)
# Technology Setup File: Morbn20.tdb
# Copyright (c) 1991-94
# Tanner Research, Inc. All rights reserved
#
****************************************************************


# connections according to fab / device physics
```

```
connect(N Well, ndiff, NotPBase)
connect(subs, pdiff, pdiff)
connect(allsubs, subs, subs)
connect(pdiff, P Base, pdiff)
connect(P Base, subs, P Base)
connect(ndiff, Metal1, ActiveContact)
connect(pdiff, Metal1, ActiveContact)
connect(poly wire, Metal1, Poly Contact)
connect(Poly2, Metal1, Poly2 Contact)
connect(Metal1, Metal2, Via)

# NMOS transistor with poly1 gate
device = MOSFET(
                RLAYER=ntran;
                Drain=ndiff, WIDTH;
                Gate=poly wire;
                Source=ndiff, WIDTH;
                Bulk=subs;
                MODEL=NMOS;
                )
# PMOS transistor with poly1 gate
device = MOSFET(
                RLAYER=ptran;
                Drain=pdiff, WIDTH;
                Gate=poly wire;
                Source=pdiff, WIDTH;
                Bulk=N Well;
                MODEL=PMOS;
                )

# NMOS transistor with Poly2 gate
device = MOSFET(
                RLAYER=ntran2;
                Drain=ndiff, WIDTH;
                Gate=Poly2;
                Source=ndiff, WIDTH;
                Bulk=subs;
                MODEL=poly2NMOS;
 )
# PMOS transistor with Poly2 gate
device = MOSFET(
                RLAYER=ptran2;
                Drain=pdiff, WIDTH;
```

```
                        Gate=Poly2;
                        Source=pdiff, WIDTH;
                        Bulk=N Well;
                        MODEL=poly2PMOS;
                        )
        #
        # NPN transistor
        device = BJT(
                    RLAYER=P Base;
                     Collector=N Well;
                    Base=P Base;
                    Emitter=ndiff;
                     Substrate=allsubs;
                     MODEL=NPN;
                    )
        #
        # Poly resistor
        device = RES(
                    RLAYER=Resistor ID;
                     Plus=poly wire, WIDTH;
                    Minus=poly wire, WIDTH;
                     MODEL=;
                     )

        # Poly-Poly2 capacitor
        device = CAP(
                     RLAYER=Capacitor ID;
                    Plus=poly wire;
                    Minus=Poly2;
                    MODEL=;
                    )

        # Bonding Area
        #device = SUBCKT(
        #                   RLAYER=Pad Comment;
        #                   Pin1=Metal1;
        #                   MODEL= PADBOND;
    #                       )
```

The syntax is described in detail in the next section. However, it is easy to identify each element and the basic layers involved in each.

11.7 Extractor Definition File

The extractor definition file is used to define devices and connections to be recognized by L-Edit/Extract.

11.7.1 Extractor Definition File Syntax and Semantics

The listing below is a formal BNF metalanguage description of the extractor definition file syntax.[1] Non-terminals appear to the left of the equal sign. Terminals are case-insensitive and are enclosed in double quotation marks. A vertical bar (|) indicates that the production on either side of the bar is permissible. Productions to the right of the equal sign are terminated by a period. Items enclosed by braces ({...}) may be repeated zero or more times. Items enclosed by brackets ([...]) may appear zero or one time.

```
extractorDefFile   = {statement}.
statement          = connectStatement | deviceStatement |
                     commentStatement | sep.
connectStatement   = " CONNECT" openDef {sepChar}
                     [commentStatement] layerName "," layerName ","
                     layerName ")".
deviceStatement    = "DEVICE" [sep] "=" {sepChar} deviceDefinition.
deviceDefinition   = mosfetDef | jfetDef | gaasfetDef | bjtDef | diodeDef |
                     subcktDef | inductorDef | capacitorDef | resistorDef.
mosfetDef          = "MOSFET" openDef recogLayer pinDef pinDef
                     pinDef pinDef modelType closeDef.
jfetDef            = "JFET" openDef recogLayer pinDef pinDef pinDef
                     modelType closeDef.
gaasfetDef         = "GAASFET" openDef recogLayer pinDef pinDef
                     pinDef modelType closeDef.
bjtDef             = "BJT" openDef recogLayer pinDef pinDef pinDef
                     pinDef modelType closeDef.
diodeDef           = "DIODE" openDef recogLayer pinDef pinDef
                     modelType closeDef.
subcktDef          = "SUBCKT" openDef recogLayer {pinDef}
                     modelType closeDef
inductorDef        = "IND" openDef recogLayer pinDef pinDef
                     modelType closeDef.
capacitorDef       = "CAP" openDef recogLayer pinDef pinDef
                     modelType closeDef.
resistorDef        = "RES" openDef recogLayer pinDef pinDef
                     modelType closeDef.
recogLayer         = {sepChar} [commentStatement] {sepChar}
```

----

[1] For more information on the BNF metalanguage, please refer to an appropriate text on formal languages.

		"RLAYER" {sepChar} "=" layerName ";".
openDef	=	[sep] "(".
closeDef	=	[sep] ")".
pinDef	=	{sepChar} pinName {sepChar} "=" layerName [widthDef] ";".
pinName	=	anyChar {anyChar}.
layerName	=	{sepChar} layerString {space}.
layerString	=	anyChar [{layerChar} anyChar].
layerChar	=	anyChar	space.
space	=	ASCII decimal 32.
widthDef	=	"," {sepChar} "WIDTH" {sepChar}.
modelType	=	{sepChar} "MODEL" {sepChar} "=" {sepChar} modelName {sepChar}.
modelName	=	anyChar {anyChar}.
sep	=	{commentStatement	sepChar}.
commentStatement	=	"#" {anyChar	sepChar} endLine.
endLine	=	Line Feed, ASCII decimal 10.
sepChar	=	any ASCII character except anyChar.
anyChar	=	upperChar	lowerChar	symbolChar	digitChar.
upperChar	=	"A"	"B"	...	"Z".
lowerChar	=	"a"	"b"	...	"z".
symbolChar	=	"!"	"""	"#"	"$"	"%"	"&"	"'"	"("	")"	
		"*"	"+"	","	"-"	"."	"/"	":"	";"	"<"	
	=	"	">"	"?"	"@"	"["	"\"	"]"	"^"	"_"	
		"`"	"{"	"	"	"}"	"~".
digitChar	=	"0"	"1"	...	"9".

11.7.2 Using Extractor Definitions

The extractor definition file contains a list of connection statements, device statements, and comments. All of these statements follow a set of rules, as outlined below:

1. The extractor definition file is case-insensitive, so both upper and lower case can be used interchangeably.

2. Whenever a layer name is listed in the file, it must match the name of a layer defined in the TDB file containing the layout to be extracted. Layer names cannot begin or end with spaces, but can have spaces contained within them. Layer names are case-sensitive.

11.7.3 Comment Statements

Comment statements begin with a pound-sign (#) and continue to the end of the line:

    # This is an extractor definition file comment.

11.7.4 Connection Statements

A connection statement instructs L-Edit/Extract how to recognize a connection between two different process layers. A connection always involves three layers: the two layers being connected, and the "via" or "contact" layer that connects them. A connection statement has the following format:

CONNECT (Layer1, Layer2, ThroughLayer)

where Layer1 and Layer2 are the names of the layers being connected, and ThroughLayer is the name of the connecting layer. Example connection statements might look like:

```
# Connect Poly to Metal1
CONNECT (Poly, Metal1, PolyContact)

# Connect Metal1 to Metal2
CONNECT (Metal1, Metal2, VIA)

# Here's a well contact
# The following layers are L-Edit"derived layers":
# Active2 = Active &(SourceDrain) &(Gate)
# SourceDrain = (ActInSel | ActInWell) &(Gate)
# ActInSel =Active & Select
# Select =(NWell) & NSelect
# ActInWell = Active & Well
# Well =(NSelect) & NWell
# Gate = Poly & Active
CONNECT (Active2, NWell, NSelect)
```

11.7.5 Device Statements

A device statement instructs L-Edit/Extract how to recognize a device. Devices are divided into two classes: passive devices and active devices. Passive devices include capacitors, inductors, and resistors. Active devices include BJTs, diodes, GaAsFETs, JFETs, MOSFETs, and subcircuits.

All devices require that a recognition layer be identified. You may specify multiple devices with the same recognition layer (with different pins), and the extractor will successfully extract these devices. This is particularly useful in extracting multi-source/drain transistors. The recognition layer is defined as follows:

RLAYER = rLayer;

where RLAYER is a required keyword and rLayer is the name of the recognition layer (corresponding to an L-Edit layer).

Following the recognition layer is a list of pins on the device. The order of this list determines the order in which L-Edit/Extract will output the pins into the extracted netlist. L-Edit/Extract does not require any particular order, but LVS requires that both source netlists contain pins in the same order, and SPICE simulators are also strict about the order in which the pins appear. For these reasons, we recommend that you follow the SPICE order of drain/gate/source/bulk or collector/

---

base/emitter/substrate.

If certain pin names are used, then L-Edit/Extract will sort the pins according to the above order. These pin names are Collector, Base, Emitter, and Substrate for BJT devices, and Drain, Gate, Source, and Bulk for all other active devices.

Pins have the following format:

pinName = pinLayer [, WIDTH];

where pinName is the name of the pin (e.g. Drain) and pinLayer is the name of the L-Edit layer associated with this pin. The WIDTH is defined as the sum of all edges of the polygon on the recognition layer that touch the pin. The length of the device is calculated by dividing the area of the device polygon by the width as computed above. Be careful! This parameter has no meaning for a gate or bulk pin, but L-Edit/Extract will not complain.

L-Edit/Extract uses the formula $R = (R_s l/w)$ for calculating the value of the extracted resistance, where R_s is the sheet resistance in units of ohms/square, w is the width, and l is the length. The value of R_s is taken from the number entered into the Setup Layer dialog box for the recognition layer of the resistor. The values of w and l are determined from the layout.

The extractor computes the area of the Recognition Layer and divides it by the effective width to obtain l. Effective width is the average of Wplus and Wminus. Note that depending on your geometry, more than two terminals may be formed; if so, the Width reported is the sum of all terminal widths divided by two. It is critical that the Width parameter be included in the extractor definition file for both terminals. A resistor with a width of 4 would be reported as having a width of 2, if the Width parameter is not included in the definition of one of the terminals.

Following the list of pins is a model definition. This definition is not required by L-Edit/Extract (i.e. MODEL=; is acceptable). The model name, if present, will be written into the extracted netlist. For SPICE, model names are not generally required for capacitors, diodes, inductors, or resistors, but are required for all other devices. Model statements have the form

MODEL = modelName;

where MODEL is a required keyword and modelName is the optional model name. Note that the MODEL=; is still required if no model name is specified. Refer to a SPICE reference manual for more information about models.

Capacitors

Capacitor statements have the form:

DEVICE = CAP (
           RLAYER = rLayer;
           Positive = pLayer;
           Negative = nLayer;
           MODEL = modelName;
)

where rLayer, pLayer and nLayer are layer names. The optional WIDTH parameter instructs L-Edit/Extract to output the dimensions of the corresponding region (pin). The modelName is the SPICE capacitor model name, typically MODEL = C.

Resistors

Resistor statements have the form:

    DEVICE = RES (
                    RLAYER = rLayer;
                    PIN1 = Layer1 [, WIDTH];
                    PIN2 = Layer2 [, WIDTH];
                    MODEL = modelName;
)

with rLayer, Layer1, and Layer2 as the layer names. The WIDTH parameter works the same as for a capacitor. The modelName is the SPICE resistor model name, typically MODEL = R.

Inductors

Inductor statements have the form:

    DEVICE = IND (
                    RLAYER = rLayer;
                    PIN1 = Layer1;
                    PIN2 = Layer2;
                    MODEL = modelName;
)

where rLayer, Layer1 and Layer2 are layer names. The WIDTH parameter works the same as for a capacitor. The modelName is the SPICE inductor model name, typically MODEL=L.

Bipolar Junction Transistors

BJT statements have the form:

    DEVICE = BJT (
                    RLAYER = rLayer;
                    Collector = cLayer [, WIDTH];
                    Base = bLayer;
                    Emitter = eLayer [, WIDTH];
                    Substrate = sLayer;
                    MODEL = modelName;
)

with rLayer, cLayer, bLayer, eLayer, and sLayer are layer names. The modelName is the SPICE model name (e.g. MODEL = NPN or MODEL = PNP).

Diodes

Diode statements have the form

    DEVICE = DIODE (
                    RLAYER = rLayer;
                    PIN1 = Layer1;

```
                              PIN2 = Layer2;
                              MODEL = modelName;
                              )
```

where rLayer, Layer1, and Layer2 are layer names. The modelName is the SPICE model name, typically MODEL=D.

GaAs MESFETs

GaAsFET statements have the form:

```
    DEVICE = GAASFET (
                          RLAYER = rLayer;
                          Drain = dLayer [, WIDTH];
                          Gate = gLayer;
                          Source = sLayer [, WIDTH];
                          Bulk = bLayer;
                          MODEL = modelName;
                          )
```

with rLayer, dLayer, gLayer, sLayer, and bLayer as layer names. The modelName is the SPICE model name.

Junction FETs

JFET statements have the form:

```
    DEVICE = JFET (
                      RLAYER = rLayer;
                      Drain = dLayer [, WIDTH];
                      Gate = gLayer;
                      Source = sLayer [, WIDTH];
                      Bulk = bLayer;
                      MODEL = modelName;
                      )
```

with rLayer, dLayer, gLayer, sLayer and bLayer are layer names. The modelName is the SPICE model name (e.g. MODEL=NJF or MODEL=PJF).

MOSFETs

MOSFET statements have the form:

```
    DEVICE = MOSFET (
                        RLAYER = rLayer;
                        Drain = dLayer [, WIDTH];
                        Gate = gLayer;
                        Source = sLayer [, WIDTH];
                        Bulk = bLayer;
                        MODEL = modelName;
                        )
```

with rLayer, dLayer, gLayer, sLayer, and bLayer as the layer names. The model-Name is the SPICE model name (e.g. MODEL = NMOS or MODEL=PMOS).

Subcircuits

Subcircuit statements have the form:

DEVICE = SUBCKT (
                  RLAYER = rLayer;
                  pinName = pinLayer [, WIDTH];
                  ...
                  MODEL = modelName;
)

List as many pin statements as needed. In this listing, rLayer and pinLayer are layer names, pinName can be any string, and modelName specifies the name of the subcircuit definition. Since a variable number of pins is permitted, with a SUBCKT definition, no pin can have MODEL as its name.

11.8 Spice Netlist Format

L-Edit/Extract supports netlist files written in a SPICE format, which is an ASCII file format intended for the interchange and device-level simulation of circuits. Tanner Tools specifically follow Berkeley 2G6 SPICE format. It is assumed that the user will have a basic familiarity with the SPICE format, or access to a SPICE reference.

11.8.1 SPICE Syntax and Semantics

Provided below is a formal BNF metalanguage description of the SPICE format written by L-Edit/Extract. Non-terminals appear to the left of the equal sign. Terminals are case-insensitive and are enclosed in double quotation marks. A vertical bar (|) indicates that the production on either side of the bar is permissible. Productions to the right of the equal sign are terminated by a period. Items enclosed by braces ({...}) may be repeated zero or more times. Items enclosed by brackets ([...]) may appear zero or one time.

spiceFile	= {spiceInstruction	deviceCall	comment} ".END" endLine.
spiceInstruction	= modelInstruction	subcktInstruction.
modelInstruction	= ".MODEL" sep modelName sep modelType.
modelType	= "C"	"D"	"L"	"NJF"	"NMOS"	"NPN" | "PJF"	"PMOS"	"PNP"	"R".
subcktInstruction	= subcktDefinition ".ENDS" endLine
subcktDefinition	= ".SUBCKT" sep subcktName sep nodeList endLine.
deviceCall	= elementCall	semiconductorCall	circuitCall.
elementCall	= cElemCall	rElemCall	lElemCall
cElemCall	= "C" deviceName sep nodeName sep nodeName [sep modelName] {sep cElemParam} endLine.
cElemParam	= capParam	multParam	genParam.
capParam	= ["C="] paramVal.
rElemCall	= "R" deviceName sep nodeName sep nodeName

	[sep modelName] {sep rElemParam} endLine.
rElemParam	= resParam I multParam I genParam.
resParam	= ["R="] paramVal.
lElemCall	= "L" deviceName sep nodeName sep nodeName
	[sep modelName] sep lElemParam endLine.
semiconductorCall	= dElemCall I qElemCall I bElemCall I
	jElemCall I mElemCall.
dElemCall	= "D" deviceName sep nodeName sep nodeName
	sep modelName {sep dElemParam} endLine.
dElemParam	= areaParam.
qElemCall	= "Q" deviceName sep nodeName sep nodeName
	sep nodeName [sep nodeName] sep
	modelName {sep qElemParam} endLine.
qElemParam	= areaParam I multParam I genParam.
areaParam	= ["AREA="] paramVal.
bElemCall	= "B" deviceName sep nodeName sep nodeName
	sep nodeName [sep nodeName] sep
	modelName {sep bElemParam} endLine.
bElemParam	= areaParam.
jElemCall	= "J" deviceName sep nodeName sep nodeName
	sep nodeName [sep nodeName] sep
	modelName {sep jElemParam} endLine.
jElemParam	= areaParam.
areaParam	= ["AREA="] paramVal.
mElemCall	= "M" deviceName sep nodeName sep nodeName
	sep nodeName [sep nodeName] sep
	modelName {sep mElemParam} endLine.
mElemParam	= lengthParam I widthParam.
lengthParam	= "L=" paramVal.
widthParam	= "W=" paramVal.
circuitCall	= "X" deviceName sep nodeName {sep
	nodeName} sep circuitName endLine.
genParamList	= genParam {sep genParam}.
deviceName	= anyName.
circuitName	= anyName.
paramVal	= value.
value	= scaleVal I expVal.
scaleVal	= posVal [scale {anyChar}].
scale	= "T" I "t" I "G" I "g" I "MEG" I "meg" I "K" I
	"k" I "M" I "m" I "U" I "u" I "N" I "n" I "P"
	I "p" I "F" I "f" I "MI" I "mi".
expVal	= posVal exp integer.
posVal	= rational I fraction.
rational	= number ["." [number]].
fraction	= "." number.
exp	= "E" I "e".

integer	=	[-] number.
number	=	anyDigit {anyDigit}.
symbolicName	=	anyName.
comment	=	commentLine	commentToEnd.
commentLine	=	"*" {anyChar} endLine.
subcktName	=	anyName.
anyName	=	anyChar {anyChar}.
sep	=	blank	commentToEnd	continueLine.
continueLine	=	blank endLine "+".
anyChar	=	upperChar	lowerChar	symbolChar	anyDigit.
upperChar	=	"A"	"B"	...	"Z".
lowerChar	=	"a"	"b"	...	"z".
symbolChar	=	"!"	"""	"#"	"$"	"%"	"&"	"'"	"(" | ")"	"*"	"+"	","	"-"	"."	"/"	":"	";"	"<"	"="	">"	"?"	"@"	"["	"\"	"]"	"^"	"_"	"'"	"{"	"|"	"}"	"~".
anyDigit	=	"0"	"1"	"2"	"3"	"4"	"5"	"6"	"7"	"8"	"9".
endLine	=	{blank} endLineChar.
endLineChar	=	ASCII end-of-line, typically CR [LF] for DOS, and LF for Unix and Macintosh.
blank	=	any ASCII character except anyChar and endLineChar.

In addition to the above syntax, the maximum number of characters per line is 80. Statements that need multiple lines can do so with the "+" continuation character as the first character on the next line.

Also, the end-of-line character or characters depends on the operating system. If you are moving a SPICE file from one operating system to another, be sure to correct for differences in the way lines are terminated.

11.9 L-Edit/Extract SPICE Commands

By using the correct commands, an entire circuit and all contained devices can be described in SPICE. Each of the commands is discussed below.

11.9.1 Passive Elements

Passive SPICE elements include capacitors, resistors, inductors, and transmission lines. Passive element statements are of the following syntax:

Cxxxn1n2[mName][C=]val

Rxxxn1n2[mName][R=]val

Lxxxn1n2[mName][L=]val

xxx is a unique name for the element; n1, n2 are names of nodes; and mName is the

---

optional model name of the device. If a model name is used, it must be defined with a .MODEL statement (see Model Statement below). Then comes a list of parametrical values for the elements, which vary by the type of element: C for capacitance, R for resistance, L for inductance. Parameters may appear in any order, with one exception: if an optional parameter name (such as the C= for a capacitor) is not specified, then that parameter must be the first parameter listed. Other unlisted parameters might appear in other SPICE documentation, but are not output by the Extractor. Refer to Parameter Values below for more information about acceptable value formats for parameters. An example capacitor statement might look like:

C120 N1997 SET1 C=120pF

This is an example of a 120pF capacitor with the unique name of C120. One pin is connected to node N1997, and the other pin is connected to node SET1.

11.9.2 Semiconductor Devices

Semiconductor devices include diodes, BJTs, GaAs devices, JFETs, and MOS-FETs. Semiconductor device statements are of the following syntax:

Dxxxn1n2mName[[AREA=]areaVal]

Qxxxncnbne[ns]mName[[AREA=]areaVal]

Bxxxndngns[nb]mName[[AREA=]areaVal]

Jxxxndngns[nb]mName[[AREA=]areaVal]

Mxxxndngns[nb]mName[L=val][W=val]

Just as with passive elements, the xxx is a unique device name, node names begin with the letter "n", and mName is the model name of the device. The model name is required for semiconductor devices, and must be defined with a .MODEL statement (see Model Statement below). Then comes a list of parametrical values for the devices, which vary by the type of device: AREA is the device area, L and W are length and width. Parameters may appear in any order. Refer to Parameter Values below for more information about acceptable value formats for parameters. An example MOSFET statement might look like:

M12 17 19 21 21 PMOS L=2U W=28U .

This is an example of a PMOS transistor with the unique name of M12. The drain node is 17, the gate node is 19, the source node and bulk nodes are the same, 21. The transistor has a length of 2 microns and a width of 28 microns.

11.9.3 Subcircuit Elements

SPICE syntax also has the capability of defining arbitrary circuits with the aid of the subcircuit function. A subcircuit consists of a list of elements and devices. A subcircuit can be called repeatedly with a subcircuit element command, which has the following syntax:

Xyyy n1 [n2] [n3] ... cName [param1=val]

+ [param2=val] ...

As with other devices and elements, yyy is a unique name, the node names are n1, n2 and n3, and cName is the subcircuit name. There must be as many node names listed as there are in the subcircuit definition. If a particular subcircuit parameter is not specified in a subcircuit element statement, its default value is assumed from the subcircuit definition statement. An example subcircuit statement would look like:

> X123 N125 N253 N74 myCircuit AREA=100 Q=42 E=17

This example is a subcircuit called myCircuit. Its unique name, to distinguish it from other instances of the same subcircuit, is X123. It has three pins, connected to nodes N125, N253 and N74. The definition for this subcircuit is given as an example in the Subcircuit Definitions section below.

11.9.4 Subcircuit Definitions

The syntax of a subcircuit definition statement is as follows:

> .SUBCKT cktname pin1 [pin2] [pin3] ... [param1=val]
>
> + [param2=val] [param3=val] ...
>
> <subcircuit statements go in here>
>
> .ENDS [cktname]

In the first line of a subcircuit definition, the cktname refers to the name, or type, of the circuit. Following the circuit name is a list of the pins (inputs and outputs) to the circuit.

The last line of the subcircuit definition can optionally contain the same subcircuit name used in the first line of the definition. Examples are:

> .ENDS
>
> .ENDS MYCIRCUIT

In between the first (.SUBCKT) and last (.ENDS) lines are any number of other SPICE statements, elements, devices, etc. The only statements not allowed within a subcircuit are other subcircuit definition statements and model statements (see Model Statement below). If no statements are inserted, then the element must be defined in an element definition file to be used with LVS.

L-Edit/Extract will not insert the body of the subcircuit between the .SUBCKT and .ENDS statement. The SUBCKT mechanism has been adopted in L-Edit/Extract primarily to aid in doing LVS verification of non-standard (i.e., non-SPICE) elements such as CCDs. If you wish to simulate your subcircuit, do not use the SUBCKT statements, or be prepared to supply an appropriate body to handle the subcircuit during simulation.

11.9.5 Model Statement

A model statement defines a model name to be used in device statements. The model statement can appear anywhere in the SPICE file, even after the model being defined is used in an element statement. The format of a model statement is as follows:

> .MODEL modelname ...

where modelname is the name of the model that is specified in the extractor definition file.

An example model statement might look like:

.MODEL MYDEVICE

Once the above statement has been issued, then a device statement such as the following could be issued:

M123 42 51 7 MYDEVICE L=2 W=28

The above statement defines a transistor with the unique name of M123. Its drain is connected to node 42, its gate to node 51, its source to node 7. It has a length of 2 and a width of 28.

11.9.6 End Statement

A SPICE file is terminated with an end statement as the last line of the file. Anything following the end statement is ignored. The syntax is as follows:

.END

11.9.7 Comments

L-Edit/Extract writes line comments. Each comment line begins with an asterisk (*).

11.9.8 L-Edit/Extract Extensions to SPICE

The SPICE format only allows for the primitive elements and devices described in the above BNF. Because a layout can contain many non-standard devices (such as multi source/drain transistors, CCDs, etc.), there is a need for a mechanism to include these in your netlist. The mechanism L-Edit/Extract uses includes these special devices as empty subcircuit definitions, with calls to the subcircuits for each instance of a non-standard device. For simulation purposes you can manually edit the subcircuit definitions.

Index

L-Edit™ Student Version

SOFTWARE LICENSE AGREEMENT
This agreement is between Tanner Research, Inc., located at 180 North Vinedo Avenue, Pasadena, California, 91107, USA, ("Tanner Research"), and you, ("Licensee"). Tanner Research grants Licensee a non-transferable, non-exclusive license to use the L-Edit Student Edition, L-Edit/DRC™ Student Edition, and L-Edit/Extract™ Student Edition programs and accompanying materials according to the following terms:

LICENSE:
You may:
a) make one (1) archival copy of the program in machine readable form for the sole purpose of backing up the software and protecting your investment from loss, provided that you reproduce all proprietary notices on the copy;
b) physically install or transfer the programs to a computer, provided that the programs are not used by more than one individual at any one time; and
c) transfer the programs onto a hard disk only for use as described above provided that you can immediately prove ownership of the original media issued by Tanner Research.

You may not:
a) use the programs in a network;
b) modify, translate, reverse engineer, decompile, disassemble, create derivative works based on, or copy (except for the archival copy) the programs or accompanying materials;
c) rent, transfer or grant any rights in the programs in any form or accompanying materials to any persons or entity without the prior written consent of Tanner Research which, if given, is subject to the conferee's consent to the terms and conditions of this license;
d) remove any proprietary notices, labels, or marks on the programs and accompanying materials; or
e) copy any written materials that accompany the programs.
This license is not for sale. Title and copyrights to the programs, accompanying materials, and any copy made by Licensee remain with Tanner Research.

TERMINATION
Unauthorized copying of the programs (alone or merged with other software) or the accompanying materials, or unauthorized transfer of the programs or license, or failure to comply with the above restrictions will result in automatic termination of the license and will make available to Tanner Research other legal remedies. Upon termination you will destroy or return the programs, accompanying materials and any copies to Tanner Research or to the location where you obtained them.

WARRANTY DISCLAIMER
NEITHER TANNER RESEARCH NOR PWS PUBLISHING COMPANY MAKE WARRANTY OR REPRESENTATION, EITHER EXPRESS OR IMPLIED, WITH RESPECT TO THESE PROGRAMS OR ACCOMPANYING MATERIALS, INCLUDING THEIR QUALITY, PERFORMANCE, MERCHANTABILITY, OR FITNESS FOR A PARTICULAR PURPOSE.

LIMITATION OF LIABILITY
IN NO EVENT WILL TANNER RESEARCH OR PWS PUBLISHING COMPANY BE LIABLE FOR DIRECT, INDIRECT, SPECIAL, INCIDENTAL OR CONSEQUENTIAL DAMAGES ARISING OUT OF THE USE OR INABILITY TO USE THE PROGRAMS OR ACCOMPANYING MATERIALS. THIS LIMITATION WILL APPLY EVEN IF TANNER RESEARCH OR AUTHORIZED DEALER HAS BEEN ADVISED OF THE POSSIBILITY OF SUCH DAMAGE. IN PARTICULAR, NEITHER TANNER RESEARCH NOR PWS PUBLISHING COMPANY IS RESPONSIBLE FOR ANY COSTS INCLUDING BUT NOT LIMITED TO THOSE INCURRED AS A RESULT OF LOSS OF USE OF THE PROVIDED PROGRAMS, LOSS OF DATA, THE COST OF A SUBSTITUTE PROGRAM, CLAIMS BY THIRD PARTIES OR FOR OTHER SIMILAR COSTS.

GENERAL
This Agreement shall be construed, interpreted, and governed by the laws of the State of California. In any dispute arising out of this Agreement, Tanner Research and you each consent to the jurisdiction of the state and federal courts of Los Angeles County, California, United States of America.
Use, duplication, or disclosure by the U.S. Government is subject to restrictions stated in paragraph (c) (i) (ii) of the Rights in Technical Data and Computer software clause 252.227.7013.
This Agreement is the entire agreement between Tanner Research and Licensee and supersedes any other communications with respect to the programs and accompanying materials. If any provision of this Agreement is held to be unenforceable, the remainder of this Agreement shall continue in full force and effect.

For damaged disk replacement only, contact:
International Thomson Publishing Customer Service

800-354-9706